JAN F. RABEK

Photodegradation of Polymers

Springer
Berlin
Heidelberg
New York
Barcelona
Budapest
Hong Kong
London
Milan
Paris
Santa Clara
Singapore
Tokyo

Jan F. Rabek

Photodegradation of Polymers

Physical Characteristics and Applications

With 94 Figures and 26 Tables

 Springer

JAN F. RABEK

Polymer Research Group
Department of Dental Biomaterials Science
Karolinska Institute
Royal Institute of Medicine
Stockholm, Sweden

ISBN 3-540-60716-1 Springer-Verlag Berlin Heidelberg New York

Cataloging-in-Publication Data applied for

Die Deutsche Bibliothek – CIP-Einheitsaufnahme
Rabek, Jan F.:
Photodegradation of polymers : physical characteristics and
applications ; with 28 tables / Jan F. Rabek. - Berlin ;
Heidelberg ; New York ; Barcelona ; Budapest ; Hong Kong ;
London ; Milan ; Paris ; Santa Clara ; Singapore ; Tokyo :
Springer, 1996
  ISBN 3–540–60716–1

Coverdesign: Lewis + Leins, Berlin
Typesetting: Scientific Publishing Services (P) Ltd., Madras
Production: PRODUserv Springer Produktions-Gesellschaft, Berlin
SPIN 10114982   2/3020-5 4 3 2 1 0 – Printed on acid-free paper

In memory of my father,
Professor T.I. Rabek, founder of the
Polymer Technology Institute
at Wrocaw Technical University, Poland,
an outstanding scientist and teacher
who introduced me to the study
of polymer chemistry

Books must follow sciences,
and not sciences books
Robert Greene (1560–1592)

Preface

The last two decades have seen dramatic advances in the understanding of the chemical reactions involved in the photodegradation of polymers. However, there are still many unanswered questions regarding the effects of UV radiation on the physical, rheological and mechanical properties of polymers.

This book was written to introduce scientists, engineers and advanced students working in polymer science and technology to the physical characteristics and practical aspects of polymer photodegradation. The changes in the structure-property relationships under UV and visible light irradiation are presented. Rather than providing a theoretical treatment of a subject only, experimentally collected and practical information on experimental work is offered. The reader can use this to complete his or her knowledge of the chemical aspects of polymer photodegradation, i.e. the mechanisms of reactions. Many issues are deliberately not considered in detail, as it has been necessary to keep this book within manageable portions (200–250 pages). The published literature on the photodegradation of polymers is vast, probably amounting to several thousand references. Only some of the most important references are mentioned to provide the reader who wishes to explore the literature in greater depth with a small number of useful starting points.

This book represents the author's experience, based on his 30 years in research on the photodegradation and photostabilization of polymers. The author has never abandoned the laboratory in that time and has maintained his enthusiasm for empirical science. It is hoped that the information provided in this book will enable the reader to become more familiar with the physical characteristics and applied aspects of polymer photodegradation in greater depth and on a broader front.

I must acknowledge my gratitude to my wife Ewelina whose patience, support and companionship have always facilitated my work, and who is firmly on my side in difficult times.

The author is grateful to Professor Otto Vogl from Polytechnic University, Brooklyn, USA for his recommendation to the Editor that allowed this book to be written.

The author would also like to express his gratitude to Professor Lars-Åke Lindén, Head of the Department of Dental Biomaterials, Karolinska Institute, Royal Academy of Medicine in Stockholm for giving him the opportunity to continue to work on polymer photodegradation. Working in a well managed and professional establishment in a friendly atmosphere is a most rewarding experience.

JAN F. RABEK
Stockholm, 1996

Contents

1   Absorption of Radiation

1.1   Radiation and Its Energy

"Optical radiation" is divided into three regions: ultraviolet (UV), visible (VIS) and infrared (IR) radiations (Fig. 1.1).

From the definition, the word "light" should be used for optical radiation perceived and evaluated by the human eye. Light is psychological, being neither purely physical nor purely psychological.

For that reason, in this book the following correct terms are used: "ultraviolet" (UV) and "light" (visible radiation), whereas for "sun radiation", consisting of many wavelengths of radiation, the commonly applied term is "sunlight radiation". Ultraviolet light, used in many publications, is an erroneous term.

The behaviour of optical radiation can be attributed either to its wave-like (quantum) character or to its corpuscular (photon) character.

The energy of a photon at a given wavelength is given by the equation

$$E = \frac{c}{\lambda} \cdot h \quad \left(h = 6.62 \times 10^{-34} \, J \, Hz^{-1} \right) \tag{1.1}$$

where c is the speed of light (3×10^{17} nm s^{-1}) and λ is the wavelength (nm).

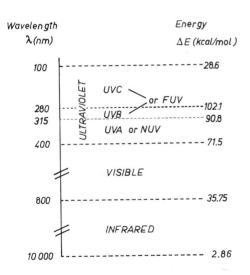

Fig. 1.1. Relationship between the wavelength of a photon and the energy (kcal mol^{-1}) of a mole of photons at that wavelength

Bond	Energy (kcal/mol)	λ (nm)
C=C	160	179
C–C	85	336
C–H	95–100	286–301
C–O	80–100	286–357
C–Cl	60–86	332–477
C–Br	45–70	408–636
O–O	35	817
O–H	85–115	249–336

Table 1.1. Energies and corresponding wavelengths for dissociation of typical chemical bonds

The energy difference ΔE (kcal mol^{-1}) for a molecule in its first excited state and in its ground state is given by

$$\Delta E = E_1 - E_0 = \frac{hc}{\lambda} = \frac{28600}{\lambda} \tag{1.2}$$

where h is Planck's constant (9.534×10^{-14} kcal s mol^{-1} or 6.626×10^{-34} Js). Figure 1.1 shows energy values calculated at different wavelengths.

The amount of energy equal to that of 1 mol of photons at a given wavelength is equivalent to the energy of 6.02217×10^{23} photons (Avogadro's number) and is called an "Einstein".

An important reference wavelength is 253.7 nm, which corresponds to the main emission line in a low-pressure mercury lamp (used very often for the study of polymer photodegradation). The equivalent energy at this wavelength is 112.7 kcal mol^{-1}. Typical average energies for homolytic cleavage of selected chemical bonds in polymers are shown in Table 1.1.

Photons at wavelengths below about 250 nm ($\Delta E = 116$ kcal mol^{-1}) possess enough energy to break most of the carbon–carbon, carbon–hydrogen, carbon–halogen, carbon–oxygen, oxygen–oxygen, and oxygen–hydrogen bonds in polymers. When polymer molecules are irradiated, however, bonds are seldom broken at random. Instead, the excited molecules undergo fairly selective bond breaking, rearrangements, or bimolecular reactions.

1.2 SI Units Used in Photochemistry

The names and symbols of SI units in photochemistry are given in Table 1.2. However, the most common presentation of photochemical data in publications is still where the energy units are expressed in kilocalories per mol. The other units, such as the joule, erg, electron-volt, or reciprocal centimeter, are mostly used in physical and theoretical publications. A conversion table for all these units is given in Table 1.3.

1.3 Absorption Cross Section

The UV (light) radiation intensity (fluence or rate incident) is defined as the number of photons crossing a given surface (n) in unit time (t):

Table 1.2. Official units in photochemistry

Physical quantity	Name of unit	Symbol
Luminous intensity	candela	cd
Work energy	joule	J
Power, radiant lux	watt	W (J/s)
Fluence	kilojoule per square meter	kJ/m^2
Fluence rate	watt per square meter	W/m^2
Luminous flux	lumen	lm (cd · sr)
Luminance	candela per square meter	cd/m^2
Illuminance	lux	lx (lm/m^2)
Wave number	reciprocal meter	m^{-1}
Radiant intensity	watt per steradian	W/sr

Table 1.3. Conversion table for units of energy

kcal/mol	eV	joule	erg	cm^{-1}
1	4.3364×10^{-2}	4.184×10^{-3}	6.9473×10^{-14}	3.4976×10^2
2.8591×10^{-3}	1.2398×10^{-4}	1.9862×10^{-9}	1.9862×10^{-16}	1
23.060	1	1.6021×10^{-5}	1.6021×10^{-12}	8065.7
1.4394×10^{13}	6.2418×10^{11}	10^{-7}	1	5.0345×10^{15}
4.184×10^{-3}	6.2418×10^{18}	1	10^7	5.0345×10^{22}

$$I = \frac{dn}{dt} \quad \text{(photons cm}^{-2}) \ . \tag{1.3}$$

(i)   In polymer research, the radiation intensity on the sample surface is usually expressed in W m^{-2} or mW cm^{-2}.

(ii) For photochemistry it is convenient to refer to a mole of photons absorbed, expressed as an Einstein (E = 6.023×10^{23}). The corresponding units of fluence are E cm^{-2}.

(iii) Fluence can also refer to the energy carried by the beam, giving what is called the "energy fluence" (energy area^{-1}) or intensity which is energy fluence per unit time (energy area^{-1} time^{-1}). These units of fluence, energy fluence, or intensity are often referred to as "dose" or exposure.

For monochromatic radiation each absorbing molecule (chromophore) exhibits a specific cross-section area (δ_a). It represents the probability that a photon will be absorbed by a molecule [377].

The intensity of radiation (amount of photons) absorbed (dI) is proportional to the number of photon-molecule collisions:

$$-dI = \delta_a I N \ dl \tag{1.4}$$

where δ_a is the cross section per molecule (cm^2); I is the radiation intensity (fluence) (photons cm^{-2}); N is the number of absorbing molecules per cubic centimeter; l is the sample thickness (cm).

This equation can easily be integrated when the concentration of absorbing molecules is independent of sample thickness (l) and radiation intensity (I). Such a condition arises when the number of photons absorbed is only an insignificant fraction of the absorbing molecules (N). Hence the integrated equation is

$$I_a = I_0 \exp(-\delta_a Nl) \ . \tag{1.5}$$

This equation can be reduced to the Lambert-Beer relationship (cf. Sect. 1.4) with the introduction of the molar concentration (M) and the extinction coefficient (ε) defined as

$$\varepsilon = \frac{N_A \delta_a}{2.303 M} \quad (l \ mol^{-1} cm^{-1}) \tag{1.6}$$

where N_A is Avogadro's number (i.e. the number of molecules in a mole; $N_A = 6.02217 \times 10^{23}$ molecules mol^{-1}).

If ε is expressed in mol l^{-1} cm^{-1}, then δ_a becomes

$$\delta_a = 2.303 \times 10^{-3} \varepsilon \ (cm^2 mol^{-1}) \ . \tag{1.7}$$

The above equation shows that δ_a is directly proportional to the extinction coefficient.

1.4 Practical Aspects of the Lambert-Beer Equation

The "Lambert-Beer equation" is defined for a monochromatic (λ) radiation as

$$\frac{I_t}{I_0} = 10^{-\varepsilon l M} = e^{-2.303 \varepsilon l M} \quad \text{or} \tag{1.8}$$

$$A = \log_{10}(I_0/I_t) = \varepsilon l M \ \text{(dimensionless)} \tag{1.9}$$

where A is called "absorbance" (sometimes "optical density" (D) or "extinction" (E) is used instead of A); I_t is the intensity of radiation transmitted through the sample solution (or sample film); I_0 is the intensity of the incident radiation; ε is the extinction coefficient (also called "molar absorptivity" in l mol^{-1} cm^{-1}); l is the sample thickness (cm); M is the molar concentration (mol l^{-1}).

In the case of a non-absorbing polymer containing internal impurities in the form of absorbing species (chromophores), both ε and M are related to the concentration of absorbing species but not to a polymer. In many such cases it is almost impossible to determine ε and M.

The intensity of radiation absorbed by the absorbing sample (in solution or by film) is given by equation

$$I_a = I_0 - I_t = I_0(1 - 10^{-A}) = I_0(1 - 10^{-\varepsilon l M}) \ . \tag{1.10}$$

Hence, if values of absorbance (A) are larger than 2, a complete absorption of radiation will occur ($I_a \approx I_0$). It is recommended to carry out photochemical reactions with concentrations of photolyte and an optical path in such a way that all the incident radiation is absorbed. However, in some cases it is convenient to use optically diluted solutions (e.g. in studies of photolysis kinetics of photoinitiators) where the value of the photolyte absorbance allows expansion of Eq. (1.10) in series and the ignoring of terms with powers of absorbance larger than 2. This approximation leads to the equation

$$I_a = I_0 \varepsilon l M \tag{1.11}$$

for absorbed radiation.

The radiation intensities (I_0, I_a and I_t) are expressed as number of photons per second or number of photons per second and square centimeters. It is also common to give the number of photons in Einsteins.

Absorption of radiation is a function of wavelength. In Fig. 1.2 an absorption of a molecule is shown, presented as $\log \varepsilon = f(\lambda)$.

When two different chromophoric groups in a given polymer absorb at the same wavelength, the individual absorbances (A_1 and A_2) contribute but the total intensity of radiation absorbed is less than the sum of the intensities of radiation absorbed individually if each were present alone in that polymer film (or polymer solution). This intensity of radiation absorbed is distributed between the two chromophoric groups (1 and 2) in the ratio of their absorbances (A_1 and A_2). The total intensity of radiation absorbed is given by equation

$$I_a = I_0[1 - 10^{-(A_1+A_2)}] \tag{1.12}$$

of which

$$I_{a(1)} = I_0[1 - 10^{-(A_1+A_2)}]\frac{A_1}{A_1 + A_2} \tag{1.13}$$

The validity of the Lambert-Beer equation is correct only for low concentration of absorbing molecules and low radiation intensities (at higher intensities two photon absorption may occur). Therefore the differential law can be expressed as

$$-dI \doteq \varepsilon_1 I M + \varepsilon_2 I^2 M \; dl \tag{1.14}$$

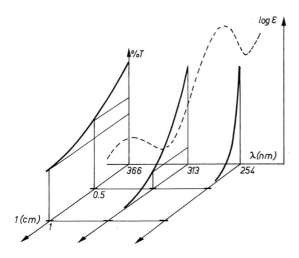

Fig. 1.2. Absorption spectrum of a molecule, $\log \varepsilon = f(\lambda)$. Fraction of light transmitted (%T) at different wavelengths as a function of the optical path (l). In this example half of incident light is absorbed within 1 cm, 1 mm and 10 μm at 366, 313 and 254 nm, respectively [241]

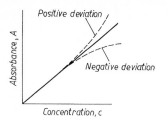

Fig. 1.3. Testing of the validity of the Lambert-Beer law, i.e. calibration curves for quantitative analysis

which, upon integration, gives

$$\frac{\varepsilon_1 + \varepsilon_2 I_t}{I_t} = \frac{\varepsilon_1 + \varepsilon_2 I_0}{I_0} 10^{-\varepsilon_1 Ml} \quad . \tag{1.15}$$

To test the Lambert-Beer equation it is necessary to measure absorbance (A) as a function of concentration (c) at a fixed wavelength (λ) and sample (cell) path (1) (Fig. 1.3). If the Lambert-Beer law is obeyed over the concentration range tested, a straight line should be obtained, passing through the origin. Deviations from the law are designated as positive or negative. Sometimes the Lambert-Beer law is obeyed at one wavelength (λ_1) but not at another (λ_2) (Fig. 1.4).

The absorption of radiation by an element of an absorbing solution (or film) is given by Eq. (1.10). As the absorption of radiation is appreciable over the cell solution (film), the absorbed intensity depends on the distance to the front of cell. This determines that the rate of the photochemical reaction, a rate that is proportional to absorbed radiation, must change from one position to another in the cell (film).

The measured rate of photochemical reaction represents an average of the local rates and is given by equation

$$\left[\frac{d[\text{product}]}{dt}\right]_{\text{measured}} = \frac{1}{1} \int_0^1 \left[\frac{d[\text{product}]}{dt}\right]_{\text{local}} dl \tag{1.16}$$

where 1 is the distance to the front of the cell (or film), and [products] is the product concentration.

In a system of complex photochemical reactions, the local rate might have a complex dependence on the absorbed intensity. For example

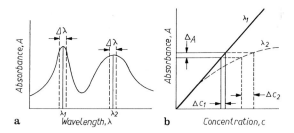

Fig. 1.4a, b. Test of the Lambert-Beer law: **a** for the absorption spectrum; **b** at wavelengths λ_1 and λ_2

$$\left[\frac{d[product]}{dt}\right]_{local} = \phi I_a \tag{1.17}$$

where ϕ is the quantum yield of a given photochemical reaction, and I_a is the intensity of radiation absorbed at the point of evaluation in the cell (or in a film).

The radiation absorbed between the distances $l + dl$ to the front of the cell (film) is given by

$$I_a = 2.303 \ \varepsilon M I_0 \ \exp(-2.303\varepsilon l M) \ . \tag{1.18}$$

The insertation of Eqs. (1.17) and (1.18) into Eq. (1.16) gives an explicit observed rate for a given photochemical reaction (e.g. photolysis):

$$\left[\frac{d[product]}{dt}\right]_{measured} = \frac{I_0(1 - \exp(-2.303\varepsilon M l))}{2.303\varepsilon M l} \ . \tag{1.19}$$

This expression can be simplified under the following two experimental conditions.

(i) The value $2.303 \ \varepsilon M l \ll 0.1$, i.e. molar concentration of absorbing species is very low (i.e. non-absorbing polymer containing internal impurities in a form of absorbing species):

$$\left[\frac{d[product]}{dt}\right]_{measured} \approx \phi \ I_0 \ 2.303 \ \varepsilon M \ . \tag{1.20}$$

(ii) The value $2.303 \ \varepsilon M l > 2$, i.e. molar concentration of absorbing species is high (i.e. polymer contains chromophoric groups as a part of molecular structure):

$$\left[\frac{d[product]}{dt}\right]_{measured} \approx \phi \frac{I_0}{l} \ . \tag{1.21}$$

However, the average absorbed intensity is given by

$$\langle I_a \rangle = \frac{\int_0^l 2.303\varepsilon M I_0 \ \exp(-2.303\varepsilon M l)dl}{\int_0^l dl} \ , \tag{1.22}$$

$$\langle I_a \rangle = \frac{I_0}{l}(1 - \exp(-2.303 \ \varepsilon M l)) \ . \tag{1.23}$$

For a highly concentrated absorbent this equation can be reduced to the form

$$\langle I_a \rangle = \frac{I_0}{l} \ . \tag{1.24}$$

Comparison of Eqs. (1.21) and (1.24) shows that the observed rate of photochemical reaction (photolysis) can now be expressed as

$$\left[\frac{d[product]}{dt}\right]_{measured} = \phi\langle I_a \rangle \ . \tag{1.25}$$

The average absorbed radiation (light) intensity $\langle I_a \rangle$ can also be used in Eq. (1.20), since for a low concentrated absorbent it is

$$\langle I_a \rangle \approx 2.303 \varepsilon M I_0 \ . \tag{1.26}$$

These equations illustrate the problem with experimental treatment of absorbed radiation (light) by very low and very high absorbing species present in common polymers.

2 Electronically Excited States in Polymers

2.1 Formation of Excited States

Photophysical processes which include formation of excited states and energy transfer processes (cf. Chap. 3) occur without change of the chemical polymer structure. The basic theory of photophysical processes has been reviewed elsewhere [62, 112, 770, 772].

An unexcited molecule in which the electron spins are paired is in the ground electronic singlet state (S_0). Absorption of a photon by the molecule results in excitation of the molecule into a new state of higher energy. If the transition occurs from a ground singlet state (S_0) without change of spin, the excited electronic state will also be a singlet (S_1). Higher excited singlets (S_2, $S_3 \ldots S_i$) will also be formed if energy of appropriate frequency is absorbed (Fig. 2.1). A plot of the intensity of absorption of the incident radiation against its frequency or wavelength gives the electronic absorption spectrum.

If the transition occurs with change of spin (two electron spins becomes unpaired) a triplet state (T_1) will be formed (Fig. 2.1). Transitions between states of unlike multiplicity are spin-forbidden and occur to a limited extent only because of spin-orbit coupling between the singlet and triplet states. Higher excited triplets (T_2, $T_3 \ldots T_i$) may be formed only when a molecule in its lowest triplet state (T_1) absorbs a new photon (triplet–triplet absorption) [491].

The energy of the first excited singlet state (S_1) is usually estimated from the wavelength at which the normalized chromophore absorption and fluorescence spectra cross [62, 112, 770, 772]. The energy of the lowest triplet state (T_1) is not easily measured. It is usually estimated from the blue edge of the low-temperature phosphorescence spectrum of the triplet chromophore [112, 317, 606, 632, 770, 772].

The excitation energy of a molecule in its excited state may be dissipated by the following processes.

(i)   Radiative processes: luminescence (fluorescence and phosphorescence).
(ii) Radiationless processes.
(iii) Bimolecular deactivation processes (energy transfer processes).
(iv) Dissociation processes, which occur when the absorption of a radiation (photon) raises a molecule from the ground state (S_0) to the repulsive excited state.

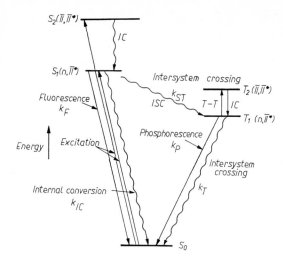

Fig. 2.1. Jablonsky diagram representing the most important photophysical processes. Radiative and radiationless transitions are indicated with straight ($\longrightarrow$) and wavy ($\rightsquigarrow$) arrows, respectively. IC is an internal conversion process, ISC is an intersystem crossing process, T–T is a triplet-triplet absorption

2.2 Radiative Processes

An electronically excited molecule can lose its excitation energy by emission of radiation, which is known as luminescence. There are two kinds of luminescence (Fig. 2.1).

(i) "Fluorescence" is a spin-allowed radiative transition between states of the same multiplicity, usually taking place from the S_1 level:

$$S_1 \longrightarrow S_0 + h\nu_F \ . \tag{2.1}$$

This emission has a short lifetime (10^{-9}–10^{-5} s).

(ii) "Phosphorescence" is a spin-forbidden radiative transition between two states of different multiplicity:

$$T_1 \longrightarrow S_0 + h\nu_P \ . \tag{2.2}$$

Phosphorescence lifetimes are about 10^{-3}–10^{-2} s.

2.3 Radiationless Transitions

"Radiationless transitions" occur between different electronic states and are induced by molecular or crystal vibrations. There are two types of radiationless transitions (Fig. 2.1) [451].

(i) "Internal conversion" (IC) is the spin-allowed radiationless transition between two states of the same multiplicity:

$$S_1 \longrightarrow S_0 + \text{heat} \tag{2.3}$$

$$T_2 \longrightarrow T_1 + \text{heat} \tag{2.4}$$

and is usually a rapid 10^{-12} s process.

(ii) "Intersystem crossing" (ISC) is the spin-forbidden radiationless transition between two states of the different multiplicity:

$$S_1 \longrightarrow T_1 + \text{heat} \tag{2.5}$$

$$T_1 \longrightarrow S_0 + \text{heat} \ . \tag{2.6}$$

The lifetime of this transition is 10^{-8}–10^{-11} s. The ISC process form S_1 to T_1 competes with fluorescence and is the most important route through which triplets are formed.

The probabilities that these processes occur in an isolated molecule are dependent on the kind of electronic transition that takes place, i.e. whether the transition is spin, symmetry, and parity allowed, and whether vibrational and spin-orbit coupling are important in the system.

The nature of the excited singlet state will certainly control the extent to which intersystem crossing to the triplet state will occur.

Polymers containing aromatic carbonyl chromophores (e.g. poly(vinyl benzophenone) have the lowest singlet state n-π^* characterized by high intersystem crossing rates and short-lived triplets.

2.4   Excitation of Chromophores

The term "chromophore" refers to the group mainly responsible for a given absorption band (Table 2.1).

Chromophores consist of π and n electrons. Absorption of radiation causes transition of π and n electrons from the ground state (S_0) to excited singlet (^1S) and triplet (^3T) states [404, 564]. Different types of reactions may occur from these excited states (Table 2.2).

Table 2.1. Typical chromophores and their spectral characteristics

Chromophore	Compound	Solvent	λ_{max}	$\varepsilon_{max}(\lambda)$
C=C	Octene-3	Hexane	185	8000
			230	2
C=O	Acetone	Hexane	188	900
			279	15
C=C—C=C	Butadiene	Hexane	217	20 900
	Benzene	Ethanol	200	4400
			256	226
	Naphthalene	Ethanol	220	110 000
			275	5600
			314	320
	Diphenyl	Chloroform	252	18 000

$\pi-\pi^*$ Singlet States
– Proton transfer reactions
– Twisting around double bonds
– Cycloaddition and cycloelimination
– Sigmatropic rearrangements
– Nucleophilic and electrophilic additions
– Cyclic rearrangements
$\pi-\pi^*$ Triplet States
– Hydrogen atom abstraction
– Addition to unsaturated bonds
– Radical rearrangements
$n-\pi^*$ States
– Atom abstraction
– Radical addition
– Electron abstraction or electron transfer
– α and β cleavages

Table 2.2. Types of excited-state reactions in organic and polymeric molecules

It is almost always the case that the lowest excited singlet state has n, π^* character. The first $^1(n, \pi^*)$ state usually lies some way above it, but the order can be reversed when the carbonyl group is bonded to a large aromatic ring system or when the molecule is dissolved in highly polar or hydrogen-bonding solvents. Because of the greater singlet-triplet separation in the π, π^* states, the lowest $^3(n, \pi^*)$ and $^3(\pi, \pi^*)$ states usually lie much closer together and their order is much more easily reversed by perturbation from the solvent or substituent groups. Bimolecular reactions of excited carbonyl chromophores such as hydrogen atom abstraction or oxetane ring formation, are typical of their $^3(n, \pi^*)$ electron states [785]:

$$\begin{array}{c} \diagdown \\ \diagup \end{array}C{=}O + PH \longrightarrow \begin{array}{c} \diagdown \\ \diagup \end{array}\overset{\cdot}{C}{-}OH + P\cdot \qquad (2.7)$$

$$^3(n,\pi^*) \qquad\qquad \text{ketyl radical}$$

$$\begin{array}{ccc} \diagdown\diagup & \diagdown\diagup & O{-}C \\ O & C & | \quad | \\ \| & \| & C{-}C \\ C + C & \longrightarrow & \diagup\diagdown\diagup\diagdown \\ \diagup\diagdown \; \diagup\diagdown & & \end{array} \qquad (2.8)$$

$$^3(n,\pi^*) \qquad\qquad \text{oxetane ring}$$

If the character of the lowest triplet level is predominantly $^3(\pi, \pi^*)$, the carbonyl group loses its biradical character and the reaction are very inefficient.

The $^3(n, \pi^*)$ state of aliphatic and aromatic ketones resembles an alkoxyl or aryloxyl radical, and if the excited molecule does not dissociate it can undergo typical free radical reactions at the electron-deficient oxygen atom (for example, inter- or intramolecular hydrogen atom transfer) [785].

The intersystem crossing (ISC) probability for the transition

$$^1(n, \pi^*) \longrightarrow {}^3(\pi, \pi^*) \quad \text{or} \tag{2.9}$$

$$^1(\pi, \pi^*) \longrightarrow {}^3(n, \pi^*) \tag{2.10}$$

is about 10^3 times as large as that for the transition between two (π, π^*) or two (n, π^*) states. The rate constants for the intersystem crossing are as follows [218]:

$$(n, \pi^*) \longleftrightarrow (\pi, \pi^*) \quad 10^9 s^{-1} \tag{2.11}$$

$$(n, \pi^*) \longleftrightarrow (n, \pi^*) \quad 10^5 s^{-1} \tag{2.12}$$

$$(\pi, \pi^*) \longleftrightarrow (\pi, \pi^*) \quad 10^6 s^{-1} . \tag{2.13}$$

2.5 Luminescence Emission from Chromophores

The luminescence (fluorescence and phosphorescence) from commercial polymers results from the presence of chromophoric groups in these polymers.

Some fluorescent and phosphorescent species, like enone and dienone(or-al) present as chromophoric impurities in commercial polyolefins are gradually consumed (disappear) during UV (or sunlight) irradiation [23, 24, 26–28, 30, 33]. This is due to the photolysis of carbonyl chromophore by the Norrish type I and type II reactions (cf. Sect. 4.23).

Thermal oxidation of polypropylene causes the appearance of the phosphoroscent ketonic/aldehyde carbonyl groups [22].

Thermal oxidation of polyamides results in the formation of fluorescent and phosphorescent species whose excitation and emission spectra closely match those of nylon 6,6 [21]. These spectra have been attributed to the excited states of α, β-unsaturated carbonyl chromophore [30, 32, 35, 36].

Polymers which contain chromophores as part of their molecular structures e.g. polystyrene, polynaphthalenes, poly(ethylene terephthalate), poly(ethersulphone), etc. show strong fluorescence and phosphorescence [29, 30].

The initial fluorescence and phosphorescence in commercial polyolefines (polyethylene and polypropylene) is mainly due to the presence of α, β unsaturated carbonyl chromophores resulting from oxidation of polymers [4, 359, 360, 551].

The great majority of carbonyl chromophores leaves the lowest excited singlet state (S_1) by the intersystem crossing transition into a neighbouring triplet level. Many are phosphorescent in a rigid medium but they rarely fluoresce, and when they do the fluorescence efficiency is usually small. Any photochemical change that follows excitation is very likely to have been initiated in the lowest triplet state.

Fluorescence spectroscopy can be used as a very rapid diagnostic technique for evaluating the extent of degradation. The intensity of the emission peaks of degradation products provides a qualitative means of establishing the level of photodegradation process [142, 631].

Fluorescence spectroscopy offers little information about individual bond properties, but otherwise has some extremely favourable features [295].

(i)   Fluorescence is the most sensitive optical spectroscopy. Useful signals can be detected at nanomolar concentrations, and at the subpicomolar level.
(ii) Fluorescence is inherently a multidimensional selective technique. The multidimensional feature, that is, the intensity depends on two wavelengths (excitation and emission).
(iii) The fluorescence excitation spectrum can be readily determined for strongly scattering or even opaque samples for which conventional UV/VIS absorbance spectroscopy is difficult or even impossible.
(iv) Fluorescence provides dynamical information on a time-scale relevant to polymer international motions.
(v)   Well-developed and well-understood models are available for data interpretations.

2.6   Kinetics of Photophysical Processes

Photophysical processes resulting from the absorption of photons by an isolated molecule in a polymer chain can be summarized as follows (Fig. 2.1):

Step		Rate	
Excitation	$S_0 + h\nu \longrightarrow S_1$	I_a	(2.14)
Internal conversion	$S_1 \rightsquigarrow S_0 + heat$	$k_{IC}[S_1]$	(2.15)
Intersystem crossing	$S_1 \rightsquigarrow T_1 + heat$	$k_{ISC}[S_1]$	(2.16)
Intersystem crossing	$T_1 \rightsquigarrow S_0 + heat$	$k_T[T_1]$	(2.17)
Fluorescence	$S_1 \longrightarrow S_0 + h\nu_F$	$k_F[S_1]$	(2.18)
Phosphorescence	$T_1 \longrightarrow S_0 + h\nu_P$	$k_P[T_1]$	(2.19)

where I_a is the rate of radiation absorption in Einsteins $l^{-1} s^{-1}$, k_{IC}, k_{ISC}, k_T, k_F and k_P are rate constants of a given process and $[S_1]$ and $[T_1]$ are concentrations of excited singlets and triplets, respectively:

$$[S_1] = \frac{I_a}{(k_{IC} + k_{ISC} + k_F)} \tag{2.20}$$

$$[T_1] = \frac{k_{ISC}[S_1]}{(k_T + k_P)} \tag{2.21}$$

or, by substituting Eq. (2.20) into Eq. (2.21),

$$[T_1] = \frac{I_a k_{ISC}}{(k_T + k_P)(k_{IC} + k_{ISC} + k_F)} \; . \tag{2.22}$$

The population of the excited singlet state (S_1) and triplet state (T_1) at any given time (t) are given by the following equations:

$$-\frac{d[S_1]}{dt} = k_{IC}[S_1] + k_{ISC}[S_1] + k_F[S_1] \tag{2.23}$$

$$-\frac{d[T_1]}{dt} = k_T[T_1] + k_P[T_1] \; . \tag{2.24}$$

Solutions of these equations are

$$\ln \frac{[S_1]}{[S_1]_0} = -(k_{IC} + k_{ISC} + k_F)t \tag{2.25}$$

$$\ln \frac{[T_1]}{[T_1]_0} = -(k_T + k_P)t \; . \tag{2.26}$$

Considering "quantum yield of fluorescence (ϕ_F)" and "phosphorescence (ϕ_P)" defined as

$$\phi_F = \frac{\text{rate of fluorescence}}{\text{rate of absorption}} = \frac{k_F[S_1]}{I} \tag{2.27}$$

$$\phi_P = \frac{\text{rate of phosphorescence}}{\text{rate of absorption}} = \frac{k_P[T_1]}{I} \tag{2.28}$$

the ratio of ϕ_P/ϕ_F is given by:

$$\frac{\phi_P}{\phi_F} = \frac{k_P}{k_F}\left(\frac{k_{ISC}}{k_T + k_P}\right) \; . \tag{2.29}$$

If the total quantum yield of emission is high, then $k_P \gg k_T$ and Eq. (2.29) becomes

$$\frac{\phi_P}{\phi_F} = \frac{k_{ISC}}{k_F} = \tau_F \, k_{ISC} \tag{2.30}$$

where τ_F is the lifetime of fluorescence.

2.7   Quenching Processes of Excited States

Excited singlet (S_1) and/or triplet (T_1) states can be deactivated by interaction of the excited molecules with the components of the system, and can be considered as bimolecular processes.

The following type of quenching processes can be distinguished:

(i)   collisional quenching (diffusion or non-diffusion controlled);
(ii)   concentration quenching;
(iii) oxygen quenching;
(iv)   energy transfer quenching;
(v)   radiative migration (self-quenching).

Quenching processes are often divided into

(i)   viscosity-dependent (dynamic) type; and
(ii)   viscosity-independent (static) type.

2.8   Quenching Processes in Solution

In the simplified kinetic consideration, the following photophysical processes can be considered in a quenching process of an excited state of molecule (D^*) by the addition of a quencher molecule (Q):

Step		Rate	
Excitation	$D_0 + h\nu \longrightarrow D^*$	I_a	(2.31)
Emission	$D^* \longrightarrow D_0 + h\nu'$	$k_E[D^*]$	(2.32)
Deactivation	$D^* \longrightarrow D_0 + heat$	$k_D[D^*]$	(2.33)
Quenching	$D^* + Q \longrightarrow D_0 + Q^*$	$k_Q[D^*][Q]$	(2.34)

where I_a is the rate of radiation absorption in Einsteins $l^{-1}\,s^{-1}$, k_E, k_D and k_Q are rate constants for emission from D^*, deactivation of D^* and quenching of D^*, and $[D^*]$ and $[Q]$ are concentrations of excited molecule D^* and quencher molecule Q.

The concentration of the excited molecule (under conditions of steady illumination and no irreversible photochemical reaction) is

$$\frac{d[D^*]}{dt} = I_a - (k_E + k_D + k_Q[Q])[D^*] \ . \tag{2.35}$$

In the steady-state condition:

$$I_a = (k_E + k_D + k_Q[Q])[D^*] \ . \tag{2.36}$$

The quantum yield for emission from an excited molecule (D^*) in the absence of quencher (Q) is given by

$$\phi_0 = \frac{k_E[D^*]}{I_a} = \frac{k_E}{k_E + k_D} \quad . \tag{2.37}$$

The quantum yield for emission from an excited molecule (D*) in the presence of quencher (Q) is given by

$$\phi_Q = \frac{k_E[D^*]}{I_a} = \frac{k_E}{k_E + k_D + k_Q[Q]} \quad . \tag{2.38}$$

Dividing Eq. (2.37) by Eq. (2.38) gives the well-known "Stern-Volmer equation":

$$\frac{\phi_0}{\phi_Q} = \frac{k_E + k_D + k_Q[Q]}{k_E + k_D} \quad \text{or} \tag{2.39}$$

$$\frac{\phi_0}{\phi_Q} = 1 + \frac{k_Q}{k_E + k_D}[Q] \quad \text{or} \tag{2.40}$$

$$\frac{\phi_0}{\phi_Q} = 1 + k_Q\tau[Q] \quad \text{where} \tag{2.41}$$

$$\tau = \frac{1}{k_E + k_D} \tag{2.42}$$

is the measured lifetime of an excited molecule (D*) in the absence of quencher molecule (Q). The plot ϕ_0/ϕ_Q vs [Q] produces a straight line with a slope of $k_Q\tau$ (Fig. 2.2).

The concentration-quenching process is conventionally described in terms of concentration $[Q]_{1/2}$ which reduces ϕ_0 to one-half of its original value (50%):

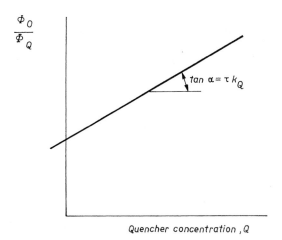

Fig. 2.2. The Stern-Volmer plot

$$\frac{\phi_0}{\phi_Q} = \frac{1}{0.5} = 2 = 1 + k_Q\tau[Q]_{1/2} \quad \text{or} \tag{2.43}$$

$$k_Q\tau[Q]_{1/2} = 1 \; . \tag{2.44}$$

Evaluation of the Stern-Volmer plot gives several important conclusions on quenching processes in solutions.

(i)   The experimental results presented in Fig. 2.3 show the quenching of the photolysis of poly(phenylvinylketone) in solution in the presence of quenchers of the excited triplet state of the aromatic ketone chromophores [462]. The slopes of the straight lines obtained allow the efficiency of various quenchers to be compared. For example, naphthalene is about twenty times more efficient as a triplet quencher when introduced into the polymer chain as vinylnaphthalene co-units than as an additive in the solution. Such a result is not unexpected since the actual concentration of the quencher inside the polymer coil is much higher than the mean quencher concentration when naphthalene is used as an additive. The process is usually diffusion controlled and k_Q is given approximately by the simplified Debye equation:

$$k_Q = \frac{8R_gT}{3000\eta} \tag{2.45}$$

where R_g is the gas constant, T is the temperature (K), and η is the liquid viscosity (in poises). For normal solvents, k_Q is calculated in this way to be about $10^{10} \, \text{l} \, \text{mol}^{-1} \, \text{s}^{-1}$.

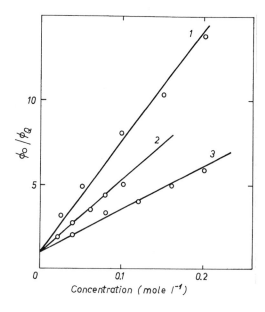

Fig. 2.3. The concentration-quenching process of poly(phenylvinyl ketone) in the presence of quenchers: (1) naphthalene; (2) 2,5-dimethyl-2,4-hexadiene; and (3) biphenyl [279, 462]

(ii) The extent to which direct interactions of excited donor polymeric car-
      bonyls with ground-state peroxides and hydroperoxides can be ascer-
      tainable by measurements of their efficiences as quenchers of the
      fluorescence intensity of the ketones [512, 515, 806].

In fact, Stern-Volmer quenching studies under steady state conditions lead to
the knowledge of the overall process, but do not allow the determination of
the states which are responsible for the quenching reaction.

2.9 Quenching Processes in Polymer Matrices

In a rigid polymer matrix, such as a solid polymer film, collisional encounter
between excited state of molecule (D*) and a quencher molecule (Q) is unlikely
through diffusion of the molecules. However, if the molecule D is the polymer
matrix, D* and Q have the chance of an encounter if interaction of an excited
molecule or group D* with identical molecules in the ground state leads to a
rapid migration of the electronic excitation (cf. Chap. 3).

    In the absence of diffusion or migration, the transfer from excited state of
molecule (D*) to a quencher molecule (Q) cannot be operative unless both
species are close together in the rigid polymer matrix. However, around each
excited molecule or group exists a "sphere of action", which is instantaneously
deactivated if there is one quencher (Q) molecule within the sphere, whereas
nothing happens if the molecule of a quencher (Q) is outside that sphere.
The probability (P) that there is no molecule of the quencher (Q) in the volume
V around the excited molecule (D*) is

$$P = \lim_{N_A[Q]\to\infty} (1 - V)^{N_A[Q]} = e^{-N_A[Q]V} \tag{2.46}$$

where N_A is Avogadro's number, [Q] is the concentration of the quencher (Q)
($mol\,l^{-1}$), and V is the "sphere of action" volume (l).

    The "yield of luminescence" originating from D* is proportional to the
probability (p) ("Perrin's equation") [570]:

$$\frac{\phi_0}{\phi_Q} = e^{N_A[Q]V} \ . \tag{2.47}$$

    The radius of the "sphere of action" (R_Q) is a measure the distance within
which the quenching becomes effective. For short range interactions, R_Q is
about 1.0–1.5 nm.

3 Electronic Energy Transfer Processes in Polymers

3.1 Energy Transfer in Photodegradation Processes

The problem of electronic energy transfer in photodegradation processes is of particular interest because a variety of energy transfer processes are possible, e.g.:

(i) initiation of photodegradation of polymers is sometimes the consequence of electronic energy transfer from an absorbing impurity to a photo-reactive group in the macromolecule;

(ii) photodecomposition of non-absorbing hydroperoxide (POOH) groups can be due to electronic energy transfer from an excited carbonyl or phenyl groups;

(iii) energy transfer processes are often responsible for the deactivation of excited chromophoric groups in polymers and in this way participate in the photostabilization of polymers.

3.2 Physical Aspects of Electronic Energy Transfer

"Electronic energy transfer process" is the one-step transfer of electronic excitation from an excited donor molecule (D^*) to an acceptor molecule (A) in separate molecules ("intermolecular energy transfer") or in different parts of the same macromolecule ("intramolecular energy transfer"):

$$D \xrightarrow{+h\nu} D^* \tag{3.1}$$

$$D^* + A \longrightarrow D + A^* . \tag{3.2}$$

Both donor (D) and acceptor (A) can be either low molecular compounds or macromolecules.

Electronic energy transfer may occur by the following mechanisms:

(I) radiative energy transfer (cf. Sect. 3.3);
(II) non-radiative energy transfer–
    (i) electron exchange energy transfer (cf. Sect. 3.4);
    (ii) resonance excitation energy transfer (cf. Sect. 3.5).

Various factors affect the extent of energy transfer between an excited donor (D^*) and the acceptor (A):

(i)   distance between D^* and A;
(ii) relative orientation to each other;
(iii) spectroscopic properties of D and A;
(iv) optical properties of a medium (polymer matrix);
(v)   effect of molecular collisions on the motion of the excited donor and an
      acceptor in the period during which the donor is excited.

In general:

(i)   the energy of the excited state (A^*) must be lower than that of D^* (the
      energy transfer process is efficient);
(ii) the sensitized excitation of A by D^* must occur within the time (τ) that
      the molecule D remains in the excited state.

3.3 Radiative Energy Transfer

"Radiative energy transfer"(the "trivial" process) of electronic energy transfer
involves the possibility of re-absorption of donor emission. The process requires two steps with an intermediate photon:

$$D \xrightarrow{+h\nu} D^* \quad \text{(excitation of donor)} \tag{3.3}$$

$$\text{I step :} \quad D^* \longrightarrow D + h\nu' \text{(emission from donor)} \tag{3.4}$$

$$\text{II step :} \quad A \xrightarrow{+h\nu'} A^* \text{(excitation of an acceptor)} . \tag{3.5}$$

No direct interaction of the donor with the acceptor is involved. Radiative
energy transfer occurs only in that region where the emission spectrum of the
donor overlaps with the absorption spectrum of the acceptor (Fig. 3.1).
The efficiency of the radiative energy transfer depends on:

(i)   quantum yield of emission from the donor;

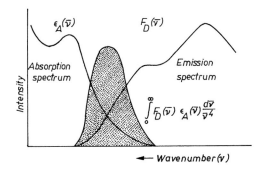

Fig. 3.1. The spectroscopic overlap integral (shaded area)

(ii) absorption of this radiation by the acceptor (governed by the Lambert-Beer law).

The radiative energy transfer may occur over very long distances (relative to molecular diameters) and the probability that an acceptor molecule reabsorbs the light emitted by a donor at a distance R varies as R^{-2}.
   The radiative energy transfer is characterized by the following.

(i)   Invariance of the donor emission lifetime. If the donor and acceptor are identical molecules, there could be a lengthening of the donor emission lifetime if multiple reasorption and re-emission occur.
(ii)   Change in the emission spectrum of the donor which can be accounted for on the basis of the acceptor absorption spectrum.
(iii) Lack of dependence of the transfer efficiency upon the viscosity of the medium.

Efficient energy transfer can take place if the donor and acceptor are different species and the spectral overlap is large. This can lead to the "internal filter phenomenon", registered as a serious distortion of the emission spectrum of donor.

3.4   Electron Exchange Energy Transfer

"Electron exchange energy transfer" occurs when an excited donor molecule (D^*) and an acceptor molecule (A) are close enough (10–15 Å) that they may be considered to be in molecular contact, i.e. their centres are separated by the sum of their molecular radii. Their electron clouds may overlap each other and an electron on D^* may also appear on A [200].
   The rate constant of electron exchange energy transfer (k_{ET}) is given by the "Dexter equation":

$$k_{ET} = \frac{2\pi K^2}{\hbar} c^{-2R/L} \int_0^\infty F_D(\bar{\nu})\varepsilon_A(\bar{\nu})d\bar{\nu} \tag{3.6}$$

where K and L are constants not available from experimental data; R is the distance between the centres of the donor and the acceptor molecules (cm), $\hbar = h/2$ is read "ħbar", h is the Planck constant, $F_D(\bar{\nu})$ is the spectral distribution of the donor emission (fluorescence intensity) in the infinitesimally small wavenumber range $\bar{\nu}$ to $(\bar{\nu} + d\bar{\nu})$ (in quanta) normalized so that

$$\int_0^\infty F_D(\bar{\nu})d\bar{\nu} = 1 \ , \tag{3.7}$$

$\varepsilon_A(\bar{\nu})$ is the molar extinction coefficient (molar absorptivity in the infinitesimally small wavenumber range $\bar{\nu}$ to $(\bar{\nu} + d\bar{\nu})$ (1 mol^{-1} cm^{-1}) normalized so that

$$\int_0^\infty \varepsilon_D(\bar{v})d\bar{v} = 1 \ , \tag{3.8}$$

and v is the wavenumber (cm^{-1}).

The spectroscopic overlap integral

$$\int_0^\infty F_D(\bar{v})\varepsilon_A(\bar{v})d\bar{v} \tag{3.9}$$

is a measure of the overlap of the donor emission and acceptor absorption.

The following spin-allowed electron exchange energy transfer processes may occur:

$$^1D^* \quad (singlet) + A \longrightarrow D + {}^1A^* \quad (singlet) \tag{3.10}$$

$$^3D^* \quad (triplet) + A \longrightarrow D + {}^1A^* \quad (singlet) \tag{3.11}$$

$$^1D^* \quad (singlet) + {}^3A^* \quad (triplet) \longrightarrow D + {}^3A^* \quad (triplet) \tag{3.12}$$

$$^3D^* \quad (triplet) + {}^3A^* \quad (triplet) \longrightarrow D + {}^1A^* \, (or {}^3A^*) \tag{3.13}$$

(triplet–triplet annihilation).

The triplet–triplet energy transfer, which is forbidden ($\times\!\!\leftrightarrow$) by the resonance-excitation energy transfer is allowed by an electron exchange energy transfer:

$$^3D^*(triplet) + A \ \times\!\!\leftrightarrow \ D + {}^3A^*(triplet) \ . \tag{3.14}$$

The energy transfer by the following route is forbidden both by an electron exchange energy transfer and the resonance-excitation energy transfer:

$$^1D^*(singlet) + A \ \times\!\!\leftrightarrow \ D + {}^3A^*(triplet) \ . \tag{3.15}$$

The electron exchange energy transfer is less important than the resonance-excitation energy transfer.

3.5 Resonance Excitation Energy Transfer

"Resonance-excitation (dipole–dipole) energy transfer"occurs when an excited donor molecule (D^*) can transfer its excitation energy to an acceptor (A) molecule over distances much greater than collisional diameters (e.g. 50–100 Å). In this mechanism, energy transfer occurs via dipole (donor)-dipole (acceptor) interaction ("Coulombic interaction"). When an acceptor (A) is in the vicinity of an excited donor (D^*) (an oscillating dipole), it causes electrostatic forces which can be exerted on the electronic systems of an acceptor [237–239].

The rate constant of the resonance-excitation energy transfer (k_{ET}) is given by the "Förster equation":

$$k_{ET} = \frac{9000\,(\ln 10)\,K^2\phi_D}{128\pi^5 n^4 N_A \tau_D R^6} \int_0^\infty F_D(\bar{v})\varepsilon_A(\bar{v})\frac{d\bar{v}}{\bar{v}^4} \qquad (3.16)$$

where K is the orientation factor for dipole–dipole interaction given by

$$K = (\cos\alpha - 3\cos\beta\cos\gamma) = 0 - 4 \qquad (3.17)$$

where α is the angle between the donor and acceptor transition moments, β is the angle between the donor moment and the line joining the centres of the donor and acceptor, and γ is the corresponding angle between the acceptor moment and the line joining the centres of the donor and acceptor.

In many polymers and biopolymers the donor and acceptor molecules undergo essentially free rotation during the lifetime of the donor. Then an average $K^2 = 2/3$ (for the random distribution of the donor and acceptor molecules) should be used.

ϕ_D is the quantum yield of the donor emission (fluorescence) in the absence of the acceptor, n is the refractive index of the medium (solvent) between the donor and the acceptor. Its value is not easy to measure when donor and acceptor are different parts of the same macromolecule. In many biopolymers n ranges from 1.3 to 1.6, N_A is Avogadro's number, τ_D is the actual mean donor (D*) emission lifetime (s); R is the distance between the centres of the donor and the acceptor molecules (cm), $F_D(\bar{v})$ is the spectral distribution of the donor emission (fluorescence intensity) defined by Eq. (3.7), and $\varepsilon_A(\bar{v})$ is the extinction coefficient (molar absorptivity) of the acceptor absorption, define by Eq. (3.8).

The Förster equation applies strictly only to those cases where:

(i)   donor and acceptor are well separated (at least 20 Å) and have restricted mobility as in polymer matrices;
(ii) the donor and acceptor exhibit broadened relatively unstructured spectra;
(iii) the spectral overlap is significant;
(iv) there are no important medium or solvent interactions;
(v)   the solvent-excited states lie much higher than those of the donor and acceptor.

The "rate constant of energy transfer"(k_{ET}) can also be presented as a function of distance between the excited donor (D*) and the acceptor (A) molecules:

$$k_{ET} = \frac{1}{\tau_D}\left(\frac{R_0}{R}\right)^6 \qquad (3.18)$$

where τ_D is the actual mean lifetime of the excited donor (D*), R is the separation between the centres of D* and A, and R_0 is the critical radius, the separation of donor and acceptor for which energy transfer from D* to A and emission from D* are equally probable (in other words the distance of separation of the donor and acceptor at which the rate of intermolecular energy transfer is equal to the sum of the rates for all other donor de-excitation processes):

$$R_0^6 = \frac{9000 \, (\ln 10) \, K^2 \phi_D}{128 \pi^5 n^4 N} \int_0^\infty F_D(\bar{v}) \varepsilon_A(\bar{v}) \frac{d\bar{v}}{\bar{v}^4} \, . \tag{3.19}$$

It is generally rather difficult to obtain the value of the quantum yield of fluorescence of the donor (ϕ_D) when donor and acceptor are different parts of the same macromolecule. One approach is to use the typical quantum yield for the donor in the model compounds. The R_0 value is not critically sensitive to uncertainties in ϕ_D because it depends only on the sixth root of ϕ_D.

The rate of energy transfer at $R_0 = R$ is equal to $1/\tau_D$ and if $R < R_0$, energy transfer dominates:

$$k_{ET} \, (\text{at } R_0) = 1/\tau_D \, . \tag{3.20}$$

The "experimental" critical radius (R_0) can be calculated from

$$R_0 = \sqrt[3]{\frac{3000}{4\pi N_A [A]_{1/2}}} = \frac{7.35}{\sqrt[3]{[A]_{1/2}}} \tag{3.21}$$

where N_A is Avogadro's number, and $[A]_{1/2}$ is the critical concentration of acceptor, the concentration at which the energy transfer is 50% efficient (emission from D^* is half-quenched) – it can be found experimentally when the deactivation of D^* in the presence of acceptor (A) is equal to one-half of its value in the absence of A, i.e

$$A = [A]_{1/2} \quad (\text{mol l}^{-1}) \, . \tag{3.22}$$

The "theoretical" critical radius (R_0) can be determined from an approximate equation (where the emission spectrum of the excited donor is expressed in terms of the absorption spectrum of the donor by using the assumed mirror-image symmetry of these spectra):

$$R_0^6 \approx \frac{k\tau_D}{\bar{v}_0^2} \int_0^\infty \varepsilon_A(\bar{v}) \varepsilon_D(\bar{v}) (2\bar{v}_0 - \bar{v}) d\bar{v} \tag{3.23}$$

where k is the constant; τ_D is the actual mean donor (D^*) emission life-time (s), $\varepsilon_A(\bar{v})$ is the extinction coefficient (molar absorptivity) of the acceptor absorption at the wavenumber $\bar{v}$ ($1 \text{mol}^{-1} \text{cm}^{-1}$), $\varepsilon_D(\bar{v})$ is the extinction coefficient of the donor absorption at the wavenumber $\bar{v}$ ($1 \text{mol}^{-1} \text{cm}^{-1}$), and $\bar{v}$ is the wavelength (wavenumber) for the O–O band in the donor spectrum (O–O donor transition).

The resonance-excitation (dipole–dipole) energy transfer probability is independent of the wavelength of the exciting radiation and increases, for a given molecule, as the extinction coefficient of the acceptor (A) and the overlap of the donor (D) emission spectrum and acceptor absorption spectrum increases.

The following spin-allowed resonance-excitation energy transfer process may occur:

$$^1D^* \quad (\text{singlet}) + A \longrightarrow D + {}^1A^* \quad (\text{singlet}) \tag{3.24}$$

$$^3D \quad (\text{triplet}) + A \longrightarrow D + {}^1A^* \quad (\text{singlet}) \; . \tag{3.25}$$

The triplet–triplet energy transfer, which is allowed by the electron exchange energy transfer, is forbidden by the resonance-excitation energy transfer:

$$^3D \quad (\text{triplet}) + A \longrightarrow D + {}^3A^* \quad (\text{triplet}) \; . \tag{3.26}$$

The singlet–triplet energy transfer is forbidden both by the resonance-excitation energy transfer and an electron-exchange energy transfer:

$$^1D^* \quad (\text{singlet}) + A \nrightarrow D + {}^3A^* \quad (\text{triplet}) \; . \tag{3.27}$$

The resonance-excitation energy transfer is more important than the electron-exchange energy transfer

3.6   Efficiency of Energy Transfer

The "efficiency of energy transfer" (ϕ_{ET}) is given by

$$\phi_{ET} = \frac{k_{ET}[A]}{k_{ET}[A] + k_D} \tag{3.28}$$

where k_{ET} is the rate constant for energy transfer; k_D is the rate constant for decay of the donor; and [A] is the concentration of the acceptor.

If k_D is a constant and independent of the concentration of acceptor (A), the efficiency of energy transfer (ϕ_{ET}) depends on the relative magnitudes of $k_{ET}[A]$ and k_D. The distance dependence of k_{ET} and hence ϕ_{ET} is quite different for the dipole–dipole and exchange energy transfer mechanisms.

The efficiency of energy transfer (ϕ_{ET}) as a function of distances between the donor (D) and the acceptor (A) is given by

$$\phi_{ET} = \frac{(R_0/R)^j}{1 + (R_0/R)^j} = \frac{(R_0/R)^6}{1 + (R_0/R)^6} \tag{3.29}$$

where R is the distance between the centres of the donor (D) and the acceptor (A) molecules (cm), and R_0 is the critical radius (cm).

The values R_0 and j can be obtained from the experimental data by plotting $\log (\phi_{ET}^{-1} - 1)$ vs $\log R$ [736]. The slope of this line is j, while R_0 is given by the value of R at $\phi_{ET} = 0.5$. In most cases $j = 6$.

The efficiency of energy transfer (ϕ_{ET}) can be determined in the following ways [519].

(i)   By measuring the quenching of donor fluorescence as a function of acceptor concentration. The energy transfer efficiency (ϕ_{ET}) is given by

$$\phi_{ET} = 1 - \frac{I_{F(D+A)}}{I_{F(D)}} \tag{3.30}$$

where $I_{F(D+A)}$ and $I_{F(D)}$ are the fluorescence intensities of the donor (D) in the presence and absence of acceptor (A) respectively; or by measuring the quenching of acceptor fluorescence as a function of donor concentration. The energy transfer efficiency (ϕ_{ET}) is given by

$$\phi_{ET} = \frac{I_{F(A)}}{I_{F(A+D)}} \tag{3.31}$$

where $I_{F(A)}$ and $I_{F(A+D)}$ are the fluorescence intensities of the acceptor (A) in the absence and presence of donor (D) respectively.

In the majority of applications these procedures are equivalent, and generally the energy transfer efficiency (ϕ_{ET}) is estimated from the donor fluorescence quenching data as no calibration of excitation sources intensity is required.

(ii) By measuring the decreased fluorescence lifetime of the excited donor in the presence of acceptor (τ_{D+A}) compared with that without the acceptor (τ_D):

$$\phi_{ET} = 1 - \frac{\tau_{D+A}}{\tau_D} \quad . \tag{3.32}$$

(iii) From the excitation spectrum corresponding to fluorescence from the acceptor, provided it is sufficiently fluorescent:

$$\phi_{ET} = \left[\frac{G(\bar{v}_2)}{G(\bar{v}_1)} - \frac{\varepsilon_D(\bar{v}_2)}{\varepsilon_A(\bar{v}_1)} \right] \left[\frac{\varepsilon_A(\bar{v}_1)}{\varepsilon_D(\bar{v}_2)} \right] \tag{3.33}$$

where $G(\bar{v})$ is the magnitude of the corrected excitation spectrum at $\bar{v}$. $G(\bar{v})$ is measured at $\bar{v}_1$, where the donor has no absorption, and at $\bar{v}_2$, where the absorption coefficient of the donor is large compared with that of the acceptor. $\varepsilon_D(\bar{v})$ and $\varepsilon_A(\bar{v})$ are corresponding absorption coefficients of the donor and acceptor, respectively.

The two first method are commonly used, whereas the third method can be used even if the local environment of the donor is different in the presence of an acceptor, provided that the absorption spectra of both the donor and acceptor groups are known.

From measurements of efficiency of energy transfer (ϕ_{ET}) it is possible to calculate the distance (R) between a donor and an acceptor molecules (cf. Eq. 3.30).

Measurements of energy transfer efficiences from studies of quenching of donor luminescence and/or enhancement of acceptor emission yield valuable information regarding relatively long-range migration effects in polymeric matrices.

From a practical standpoint, knowing the efficiencies of specific energy transfer is generally more important that knowing their inherent rates. The number of energy transfer events per donor lifetime $(k_{ET} \tau_D)$ is the critical parameter for determining energy transfer efficiency. Thus, in order to anticipate the efficiency of energy transfer both k_{ET} and τ_D must be evaluable.

A slow rate of transfer may be efficient if τ_D is long, i.e. although the probability of energy transfer per unit time is small, the period of time available to achieve energy transfer is large.

3.7   Spectroscopic Overlap Integral

The "spectroscopic overlap integral"

$$\int_0^\infty F_D(\bar{\nu}) \varepsilon_A(\bar{\nu}) \frac{d\bar{\nu}}{\bar{\nu}^4} \tag{3.34}$$

is large if the emission of an excited donor (D^*) overlaps strongly with the absorption spectrum of an acceptor (A) (Fig. 3.1). The spectroscopic overlap integral can be determined by graphical integration (Fig. 3.1). It can also be calculated from Gaussian distributions of the absorption and emission bands:

$$\varepsilon_A(\bar{\nu}) = \varepsilon_{A_{max}} \exp[-\{(\bar{\nu} - \nu_A)/\delta_A\}^2] \tag{3.35}$$

$$F_D(\bar{\nu}) = \varepsilon_{D_{max}} \exp[-\{(\bar{\nu} - \bar{\nu}_D)/\delta_D\}^2] \tag{3.36}$$

where $\varepsilon_{D_{max}}$ and $\varepsilon_{A_{max}}$ are maximum molar absorption coefficients of the longest wavelength absorption bands of the donor and acceptor molecules, respectively, $\bar{\nu}_D$ and $\bar{\nu}_A$ are wavenumber maxima of the donor and the acceptor absorption spectra, and δ_D and δ_A are standard deviations, which can be calculated by considering the overlapping sides of the absorption and emission bands (Fig. 3.1).

With the Gaussian approximations the spectroscopic overlap integral is given by

$$J = \varepsilon_{A_{max}} \varepsilon_{D_{max}} \frac{\sqrt{\pi}}{\left[\left(\frac{1}{\delta_A^2}\right) + \left(\frac{1}{\delta_D^2}\right)\right]} \exp\left[\frac{-(\bar{\nu}_A - \bar{\nu}_D)^2}{\delta_A^2 + \delta_D^2}\right]. \tag{3.37}$$

Overall probability of the energy transfer is proportional to the spectroscopic overlap integral. In the case of energy transfer between equivalent molecules, the spectroscopic overlap integral represents the extent of overlap of the emission and absorption of the donor molecule (D). In general this overlap is most favourable for chromophores that undergo minor geometric changes in the excited state.

3.8   Energy Transfer Processes in Solution

Electronic energy transfer processes in solution may occur in the following ways.

(i)   Radiative energy transfer (cf. Sect. 3.3).

Table 3.1. Characteristics of electronic transfer processes

Experimentally measured donor (D) characteristic	Radiative energy transfer	Electron exchange energy transfer	Resonance excitation energy transfer	Energy migration
Absorption spectrum	Unchanged	Unchanged	Unchanged	
Emission spectrum	Changed	Unchanged	Unchanged	Small change depending on magnitude of donar –donor interaction
Lifetime	Unchanged	Decreased	Decreased	Decreased
Dependency of rate on increasing viscosity	None	Decreased	Decreased	

(ii) Collisional transfer via electron exchange energy transfer which requires close approach (of the order of collisional diameters) (cf. Sect. 3.4).

(iii) Long-range, single-step radiationless transfer via dipole–dipole interaction (resonance-excitation energy transfer) over distances large compared to molecular diameters (cf. Sect. 3.5).

(iv) Excitation migration among donor (solvent) molecules, followed by transfer to acceptor (solute).

Experimental differentiation of these electronic transfer processes is shown in Table 3.1.

3.9 Diffusion-Controlled Reactions

In "diffusion-controlled" reactions, the rate is not dependent on the reactivity of species, but on the frequency of encounters of the reagents (i.e. their diffusion in a polymer matrix or in a solution).

The rate of diffusion is inversely proportional to the viscosity of the medium. The diffusion constant for a spherical particle (D) is given by Stoke's law:

$$D = \frac{kT}{6\pi\eta R} \qquad (3.38)$$

where k is the Boltzmann constant; T is the absolute temperature (K), η is the viscosity of medium (poise); and R is the particle radius.

The rate constant for a diffusion-controlled reaction (k_{diff}) is given by the modified Debye equation [177]:

$$k_{diff} = \frac{1}{4}\left(2 + \frac{d_1}{d_2} + \frac{d_2}{d_1}\right)\frac{8 R_g T}{3000\eta} \quad (l\,mol^{-1}s^{-1}) \qquad (3.39)$$

where R_g is the gas constant ($8.31 \times 10^7\,erg\,mol^{-1}\,deg^{-1}$), and d_1 and d_2 are diameters of the reacting species (which are assumed to be spherical).

The Debye equation can be used to determine diffusion constants at experimental temperatures if the value of D is known at some other temperature T_1. This is done by measuring the viscosity of the solvent at T_1 and the experimental temperature T_2. The required value of D_2 is given by

$$D_2 = D_1 \left(\frac{T_2}{T_1}\right) \left(\frac{\eta_1}{\eta_2}\right) \tag{3.40}$$

where η_1 and η_2 are the solvent viscosities (poise) at $T_1(K)$ and $T_2(K)$ respectively.

For species having similar dimensions, Eq. (3.39) takes the form of the simplified Debye equation:

$$k_{diff} = \frac{8 R_g T}{3000\eta} = 2.2 \times 10^5 \left(\frac{T}{\eta}\right). \tag{3.41}$$

The assumption of equivalent size of the diffusing species is no longer valid if one of the two is a polymer molecule. In this case a modified form of the "Smoluchowski equation" has to be used [705]:

$$k_{diff} = \frac{4\pi (R_1 + R_2) N_A}{1000} (D_1 + D_2) \tag{3.42}$$

Table 3.2. Diffusion-controlled rate constants (k_{diff}) and viscosities of different solvents [112]

Solvent	$k_{diff} \times 10^{-10}$, liters/mole-sec, at 20°	Viscosity, $\eta \times 10^3$, poise, at a Temperature, °C, of				
		0°	10°	20°	30°	40°
Isopentane	2.9	2.78	2.49	2.25	2.05	–
n-Pentane	2.8	2.79	2.55	2.35	2.16	–
n-Hexane	2.1	3.81	3.62	3.13	2.85	2.62
n-Heptane	1.6	5.26	4.67	4.18	3.77	3.43
n-Octane	1.2	7.13	6.20	5.47	4.87	4.37
Chloroform	1.2	6.99	6.25	5.63	5.10	4.64
Toluene	1.1	7.73	6.70	5.87	5.20	4.65
Methanol	1.1	8.08	6.90	5.93	5.15	4.49
Benzene	1.0	–	7.60	6.49	5.62	4.92
Carbon tetrachloride	0.67	14.03	11.35	9.69	8.42	7.38
Cyclohexane	0.66	–	11.80	9.80	8.26	7.04
Water	0.64	17.93	13.10	10.09	8.00	6.54
Ethanol	0.55	–	–	11.92	–	–
n-Propyl alcohol	0.29	38.6	29.2	22.6	17.7	14.0
Isopropyl alcohol	0.27	45.6	32.5	23.7	17.7	12.9
Ethylene glycol	0.038	–	–	173	–	–
Glycerol	0.00061	–	–	10700	3800	–

where $(R_1 + R_2)$ is the sum of the radii of the colliding groups, D_1 and D_2 are the corresponding diffusion coefficients, and N_A is Avogadro's number.

The rate of a diffusion-controlled reaction can be changed significantly by changing the viscosity of the solvent (cf. Eq. 3.41). In Table 3.2 are shown the viscosities of a number of common solvents and the calculated values of diffusion-controlled rate constants (k_{diff}).

The direct measurement of the diffusion coefficient of radical intermediates in solution can be obtained by using a technique of photochemical space intermittency [534, 662]. This technique is based on the facts that if photochemically produced radicals are destroyed by a second-order process, and if a cell is illuminated with a pattern of light and dark areas, the average concentration of radicals in space is dependent on the size of the light areas at a given total absorbed radiation energy. The effect is very similar in origin to that of the rotating sector technique.

For non-viscous fluids, the diffusion coefficient (D) for a small molecule is usually of the order 10^{-5} cm^2 s^{-1}, while in a viscous or solid medium, D can be of the order of 10^{-10} cm^2 s^{-1}.

3.10   Solvent Effects

The solvent may influence the rate or efficiency of an energy transfer in the following ways.

(i)   Viscosity effect (cf. Sect. 3.11).
(ii) Solvent effects on the energy levels of the donor and acceptor (absorption spectra and the overlap integral).
(iii) Solvent effect on the excited-donor lifetime (e.g. a .reversible photochemical reaction between excited donor and solvent, such as a proton transfer, which leads to quenching of the donor).
(iv) The solubility of the donor and acceptor. Good solvent provides a random distribution of donor and acceptor molecules, whereas a poor solvent might lead to a non-statistical distribution with clumping together of donor molecules, acceptor molecules, or donor and acceptor molecules. The non-statistical distribution could lead to anomalously high or low electronic energy transfer rates and efficiences.
(iv) Polarity-effects. The field in a dielectric solvent has an effect on long-range resonance-excitation (dipole–dipole) energy transfer.

There is an exceptional mechanism in which electronic energy transfer from an excited donor molecule (D^*) to an acceptor molecule (A) occurs via a series of transfer initiated by energy transfer from D^* to a solvent molecule, mediated by hopping or migration of the excitation energy through the solvent and terminated by energy transfer from a solvent molecule to an acceptor molecule (A) (Fig. 3.2) [771, 772]. In this case the solvent serves as an "electronic energy conductor".

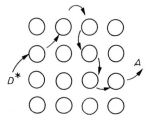

Fig. 3.2. Migration of electronic excitation via hops through the solvent

3.11 Viscosity Effects

Molecular diffusion in solution or fluid media can be depicted schematically as the relative motion of excited donor D^* and acceptor (A) molecules through the empty space between solvent molecules (or macromolecules – in this case a free volume plays an important role, cf. Sect. 5.1) (Fig. 3.3).

    To gain an appreciation of distance-time relations for molecular diffusion (or electronic energy transfer in solution), a plot of the distance (r) an excited donor molecule (D^*) will diffuse in a time period (τ) is shown in Fig. 3.4.

    The relationship between the distance (R) and time (τ) is given by the equation

$$R = \sqrt{2D\tau} \tag{3.43}$$

where D is the diffusion coefficient.

    In Fig. 3.4, Eq. (3.43) is plotted for $D = 10^{-5}\,\text{cm}^2\,\text{s}^{-1} = 10^{11}\,\text{Å}^2\,\text{s}^{-1}$, a value typical of a molecule diffusing in a fluid organic solvent, and $D = 10^{-10}\,\text{cm}^2\,\text{s}^{-1} = 10^6\,\text{Å}^2\,\text{s}^{-1}$, a value typical of very viscous, nearly rigid solvent (e.g. polymer matrix) [771]. Considering 1 ns as typical of the lifetime of a donor molecule (D^*) in the excited singlet state (s_1), this molecule in the fluid solvent will diffuse roughly 15 Å during its lifetime. In the more viscous environment, the same molecule will only diffuse about 10^{-1} Å. On other hand, a donor molecule in the triplet excited state (T_1) whose lifetime is 10^{-3} s may, during its lifetime, diffuse up to 15 000 Å in the fluid solvent or upto 50 Å in the viscous solvent.

    If every collision in solution between excited donor (D^*) and acceptor (A) molecules leads to energy transfer, the transfer rate will be a diffusion-controlled rate. That is, the rate constant of energy transfer (k_{ET}) is governed by the rate of diffusion of excited donor and acceptor molecules, i.e. $k_{ET} \approx k_{diff}$.

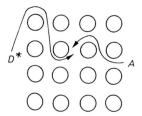

Fig. 3.3. Mutual molecular diffusion of an excited donor (D^*) and an acceptor (A) molecules

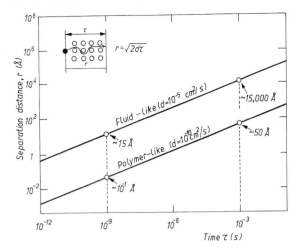

Fig. 3.4. Mean molecular displacement of a molecule in a non-viscous fluid ($d = 10^{-5}\,cm^2s^{-1}$) and a viscous fluid ($d = 10^{-10}\,cm^2s^{-1}$) [771]

The efficiency of energy transfer in solution (ϕ_{ET}) can be obtained from the simplified equation

$$\phi_{ET} = \frac{k_{ET}}{k_{diff}} = \frac{k_Q}{k_{diff}} \tag{3.44}$$

where k_{ET} is the rate constant of energy transfer and k_Q is the rate constant of quenching.

Diffusion-controlled quenching of small molecules by polymers in solution shows that quenching rate per unit is smaller than that for a model compound of low molecular weight [220, 221, 546, 580]. The quenching rate constants decrease when the molecular weight of polymers used as quenchers increases. This result allows for the determination of quenching of macromolecular volume from quenching rate constants. This quenching volume can be related to the actual volume of the macromolecular coil and to the volume of the equivalent hydrodynamic sphere.

In general, triplet energy transfer is diffusion-controlled with a unit collision efficiency, so long as the transfer is exothermic. Its efficiency may also be sensitive to the relative orientations of the donor and acceptor.

3.12   Energy Transfer in Rigid Polymer Matrix

In a polymer matrix positions of donor (D) and acceptor (A) are rigid in space (absence of molecular diffusion). With both D* and A can be associated "collisional radius" R_D and R_A which is related to the size of the molecule. Three situations with respect to a D*-A pair are possible (Fig. 3.5) [771]:

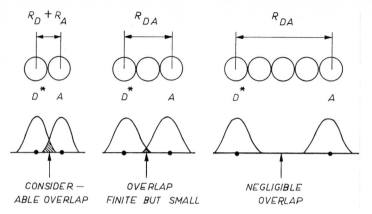

Fig. 3.5. Schematic of the overlap of electron clouds of molecules. The circles represent the van der Waals (or collisional) sizes of the molecules. These sizes underestimate the actual extension of the electron clouds in space. Below the van der Waals sizes a qualitative probability function for finding an electron is plotted as a function of nuclear electron distance [711]

(i) D^* and A are separated by distances comparable to the sum of $R_D + R_A$, i.e. molecules are undergoing collisions capable of inducing chemical interactions;

(ii) D^* and A are separated by distances of the order of twice the sum of $R_D + R_A$, i.e. molecules are incapable of undergoing strong chemical interactions, but the overlap of their electronic wave functions is still finite;

(iii) D^* and A are separated by distance of the order of several times the sum of $R_D + R_A$, i.e. the overlap of the electron clouds of the molecules is negligible.

The rate of energy transfer (k_{ET}) of the dipole–dipole interaction falls off as R_{DA}^{-6} whereas the rate of the exchange interaction falls off as $\exp - R_{DA}$. A qualitative comparison of how $k_{ET}\tau_D$ falls off for these two interactions is shown in Fig. 3.6. The efficiency of energy transfer by the exchange me-

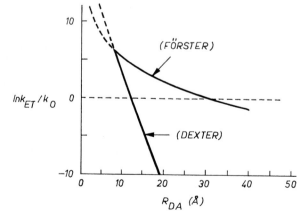

Fig. 3.6. Qualitative relationship between the ratio of the energy transfer rate constant (k_{ET} to acceptor decay constant (K_0 for Coulombic and electron exchange energy transfer to the separation between D^* and A [771]

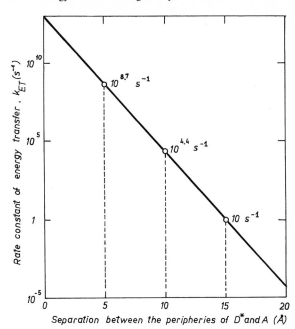

Fig.3.7. Fall of the rate constant for energy transfer by the electron exchange mechanism as a function of separation of donor and acceptor for $k_{ET} = 10^{13}$ exp 2R [771]

chanism falls off much more sharply as a function of increasing separation of donor and acceptor than the efficiency of energy transfer by the dipole–dipole mechanism.

The rate constant of energy transfer (k_{ET}) is given by the equation

$$k_{ET} = k_0 \exp - R \tag{3.45}$$

where k_0 is the maximum rate constant for energy transfer which occurs when D^* and A are in the state of a "classical" collision ($R_D + R_A = R_{DA}$) (the maximum value of $k_0 = 10^{13}$ s^{-1}), and R is the separation between the peripheries of D^* and A when they are further apart than the sum of their classical radii, i.e.

$$R = R_{DA} - (R_A + R_D). \tag{3.46}$$

In Fig. 3.7 a plot of k_{ET} vs R is shown. The value of k_{ET} falls from 10^{13} s^{-1} when D^* and A collide ($R = 0$) to $\sim 10^4$ s^{-1} when $R = 10$ Å. When D^* and A are separated by several times their collisional diameters (~ 20 Å) the value of k_{ET} is dropped by a factor of $\sim 10^{-13}$. The exchange mechanism will be generally inoperative for this magnitude of separation.

The efficiency of energy transfer arises:

(i)   when R_{DA} is of the order of $R_D + R_A$, both the electron exchange and resonance excitation (dipole–dipole) mechanisms may be effective;

(ii) when R_{DA} is much larger than $R_D + R_A$ only the dipole–dipole mechanism may be effective, but only when the extinction coefficient of an acceptor $\varepsilon_{A(max)} \sim 10^5$.

(iii) when $\varepsilon_{(max)}$ is < 1, the dipole–dipole mechanism may be ineffective even at small separation of D* and A.

3.13 Spectroscopic Methods for the Determination of Electronic Energy Transfer Processes

When an electronic energy transfer process is determined by spectroscopic methods the donor (D) is usually excited by radiation not absorbed by the acceptor (A), and then spectroscopic evidence for the formation of an excited acceptor (A*) is sought.

In order to establish that the resonance–excitation (dipole–dipole) energy transfer mechanism is operating, it must be shown that, since resonance excitation does not require collisions, it:

(i)   must be able to occur efficiently over distances considerably greater than molecular diameters; and
(ii) must be insensitive to the viscosity of the solvents or medium.

Complex formation and the radiative energy transfer (the "trivial" process) mechanism must also be carefully eliminated as the mechanism responsible for an electronic energy transfer process.

Quantitative information concerning the rates (k_{ET}) and efficiences (ϕ_{ET}) of energy transfer processes can be obtained by the fitting of k_{ET} and ϕ_{ET} equations to experimental data and then evaluating for the desired quantities.

There are two models for quantitative handling of experimental data. The first, the "Stern-Volmer model" assumes that if

(i)   the rate constant for energy transfer (k_{ET}) is independent on concentrations of excited donor (D*) and acceptor (A), and
(ii) "statistical mixing" of donor and acceptor is completely achieved during the lifetime of an excited donor (D*),

then the rate constant for decay of D* in the presence of an acceptor (A) is given by the equation

$$k_{(D+A)} = k_{(D)} + k_{ET}[A] \qquad (3.47)$$

where $k_{(D)}$ is the rate constant for decay of an excited donor D* in the absence of an acceptor (A), and [A] is the concentration of an acceptor (A). In terms of quantum yields of fluorescence emission from an excited donor (D*):

$$\frac{\phi_{F(D)}}{\phi_{F(D+A)}} = \frac{1 + k_{ET}[A]}{k_{(D)}} = 1 + k_{ET}[A]\tau_{(D)} \qquad (3.48)$$

where $\phi_{F(D)}$ and $\phi_{F(D+A)}$ are the quantum yields of fluorescence emission in the absence of A, and in the presence of A, respectively, and $\tau_{(D)}$ is the lifetime of an excited donor (D*) in the absence of A:

$$k_{(D)} = 1/\tau_{(D)} \qquad (3.49)$$

If an experimental plot of $\phi_{F(D)}/\phi_{F(D+A)}$ vs [A] yields a straight line of intercept equal to 1.0, then the "Stern-Volmer Rate Law" is fitted and the slope of the line gives the value of $k_{ET}\tau_{(D)}$. Since $\tau_{(D)}$ can be measured by independent observation, the magnitude of k_{ET} may be evaluated explicitly. The record, the "Perrin model" assumes that if

(i)   around each of the excited donor (D^*) exist an "active volume" (cross section),
(ii) if an acceptor molecule (A) is within the sphere of an active volume, D^* transfers energy to A with unit efficiency,
(iii) if an acceptor molecule (A) is outside of the sphere there is no energy transfer, and
(iv) neither molecular diffusion nor energy migration is possible, then

$$\frac{\phi_{F(D)}}{\phi_{F(D+A)}} = \exp N_A V[A] \tag{3.50}$$

where N_A is Avogadro's number, and V is the volume of the "active sphere" of energy transfer about D^*.

If a plot of $\ln \phi_{F(D)}/\phi_{F(D+A)}$ vs [A] yields a straight line, then the "Perrin Rate Law" is fitted and the slope of the line gives the value of NAV, and hence V may be evaluated. Instead of volume V, the critical radius (R_0) can be used:

$$R_0 \approx 7[A]^{-1/3}(\text{Å}) \tag{3.51}$$

where [A] is the concentration of an acceptor (mol l^{-1}). A plot of R_0 vs $[A]^{-1/3}$ is shown in Fig. 3.8.

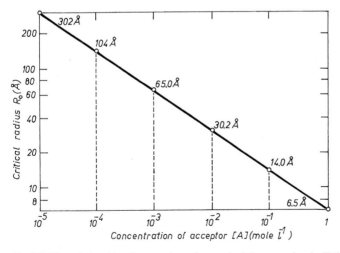

Fig. 3.8. The relationship of separation of an excited donor molecule (D^*) and acceptor molecule (A) to the concentration of [A] [771]

However, to determine whether energy transfer is occurring via electron exchange energy transfer or resonance–excitation (dipole–dipole) energy transfer, a more detailed theoretical formulation is required.

If the critical radius (R_0) determined for a solid polymer is larger than ~ 15 Å, it may be concluded that energy migration is operative (and quantitative evaluation of the energy transfer process is generally not possible).

3.14   Singlet–Singlet Energy Transfer in Polymers

"Singlet–singlet energy transfer" between an excited donor (D^*) and an acceptor (A)

$$^1D^*(S_1) + A(S_0) \longrightarrow D(S_0) + {}^1A^*(S_1) \tag{3.52}$$

may occur via:

(i)   an electron exchange interaction favoured by a small value of $k_{F(D)}$ and $\varepsilon_{A_{max}}$;
(ii) a dipole–dipole interaction (resonance excitation energy transfer) which is favoured by large value of $k_{F(D)}$ and $\varepsilon_{A_{max}}$;
(iii) an indirect mechanism involving energy migration through polymer segments (exchange or dipole–dipole interactions).

The excited donor (D^*) can be an isolated excited molecule or an excimer (cf. Sect. 3.19).

Investigations of the singlet–singlet energy transfer processes in polymers have been mainly concerned with polymers containing aromatic chromophores (Table 3.3).

3.15   Triplet–Triplet Energy Transfer in Polymers

"Triplet–triplet energy transfer" between an excited donor (D^*) and an acceptor (A) [283]

$$^3D^*(T_1) + A(S_0) \longrightarrow D(S_0) + {}^3A^*(T_1) \tag{3.53}$$

Table 3.3. Singlet–singlet energy transfer in polymeric systems

Donor	Acceptor	R_0	K_{ET}	References
Polystyrene	Tetraphenyl-butadiene	20	3×10^9	64
Poly(vinylcarbazole)	Benzophenone	26	–	282
Poly(vinylnaphthalene)	Benzophenone	15	1×10^9	167
Poly(vinylmethylketone)	Benzophenone	8	10^9	170

is "forbidden" by the dipole–dipole interaction (resonance excitation energy transfer) (exceedingly low ε_A), but is "spin-allowed" by the electron exchange (values of R_0 of the order of 10–15 Å). This mechanism requires close contact between donor and acceptor molecules. Some examples of triplet-triplet energy transfer in polymeric systems are given in Table 3.4.

The efficiency of triplet-triplet energy transfer decreases when the triplet energy of the donor approaches that of the acceptor, and falls to zero when triplet energy transfer is endothermic.

Table 3.4. Triplet–triplet energy transfer in polymeric systems

Donor	Acceptor	R_0	k_{ET}	References
Poly(vinylbenzophenone)	Naphthalene	36	10^5	165
Poly(phenylvinylketone)	Naphthalene	26		166
Poly(methylvinylketone)	Naphthalene	11	–	170
Poly(vinylnapthalene)	1,3-Pentadiene	15	10^2	168
Styrene-co-vinylbenzo-phenone	Naphthalene	300	–	169

3.16 Energy Transfer from Carbonyl Groups to Hydroperoxides

The most probable mechanism of photodecomposition of polymeric hydroperoxide (POOH) groups occurs via an energy transfer process from the excited carbonyl groups (donors) to hydroperoxy groups (OOH) (acceptors) [281, 447, 512, 733, 791, 806]:

$$
\underset{(\text{donor})}{\text{\Large$>$}C{=}O^* } + \underset{(\text{acceptor})}{P{-}OOH} \longrightarrow \underset{\text{exciplex}}{\left[{\text{\Large$>$}}C{=}O\cdots(P{-}OOH) \right]^*} \qquad (3.54)
$$

The hydroperoxide is decomposed through the intermediary of an exciplex, in which electronic energy from the CO triplet state is transferred to the vibrational states of the hydroperoxide group, resulting in the scission of the O–O bond:

$$
\left[{\text{\Large$>$}}C{=}O\cdots(P{-}OOH) \right]^* \longrightarrow P{-}O\cdot + \cdot OH + {\text{\Large$>$}}C{=}O \ . \qquad (3.55)
$$

This reaction occurs during the photo-oxidative degradation of polystyrene [281, 791, 792]. Direct absorption of UV radiation by hydroperoxides is not important since their absorption coefficients are not higher than that of polystyrene. The extinction coefficient of *tert* -butyl hydroperoxide, $\varepsilon_{260\ nm} = 6$ $1\ mol^{-1}\ cm^{-1}$ and cumene 2 hydroperoxide $\varepsilon_{260\ nm} = 200\ 1\ mol^{-1} cm^{-1}$.

Experimental data obtained for the rate of disappearance of hydroperoxy groups during the ketone-sensitized photolysis of *tert*-butylhydroperoxide and of *cis*-polyisoprene hydroperoxide suggest that a significant portion of energy is transferred from the excited carbonyl groups to the hydroperoxide groups results in direct scission of the peroxy linkage by an energy transfer mechanism [314, 315]:

$$\!\!\!\!\!\!\!\!\!\!\text{\Large$>$}\!\!\!\text{C}\!\!=\!\!\text{O}^* + \text{P}\!-\!\text{OOH} \longrightarrow \text{P}\!-\!\text{O}\cdot + \cdot\text{OH} + \text{\Large$>$}\!\!\!\text{C}\!\!=\!\!\text{O} \qquad\qquad (3.56)$$

3.17   Energy Transfer from Phenyl Groups to Hydroperoxides

Phenyl chromophores which form a part of the molecular structure of many vinyl aromatic polymers and copolymers, e.g. polystyrene and its copolymers, may transfer their excitation energy directly to hydroperoxide groups resulting from photo-oxidative degradation of these polymers [279, 281, 283, 284, 281].

In this mechanism localization of hydroperoxy groups in a polymer matrix plays an important role. In consequence, hydroperoxides can be divided into two groups [90–93, 96]:

(i)   active hydroperoxides which are formed in the neighbourhood of phenyl groups (a sphere surrounding the phenyl groups (Fig. 3.9)) photolysed by an energy transfer process in a collision complex formed between excited phenyl groups (donors) and vibrationally excited hydroperoxides (acceptors);

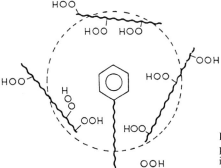

Fig. 3.9. A sphere in which energy transfer process between active hydroperoxides formed in the neighbourhood of a phenyl group may occur

(ii) inactive hydroperoxides resulting from the propagation of polymer peroxy radicals outside a sphere, which avoids photodecomposition.

After an induction period, at the beginning of irradiation, the active hydroperoxides reach a steady concentration, while the inactive hydroperoxides continue to increase slowly, but regularly, with irradiation time.

3.18 Excited State Annihilation Processes

There are two types of excited "annihilation processes" –

(i) homogenous annihilation

$$A^* + A^* \longrightarrow A^{**} + A_0 \quad \text{and} \tag{3.57}$$

(ii) heterogenous annihilation

$$A^* + B^* \begin{cases} \nearrow A^{**} + B_0 & (3.58) \\ \searrow A_0 + B^{**} & (3.59) \end{cases}$$

where A^* or B^* are molecules (or chromophores in excited singlet (S_1) or triplet (T_1) states, and A^{**} or B^{**} are molecules (or chromophores) in highly excited singlet (S_2, $S_3 \ldots S_i$) or triplet (T_2, $T_3 \ldots T_i$) states. In these processes the highly excited state (A^{**} or B^{**}) may rapidly degrade to the lowest excited state of that multiplicity, or may undergo a chemical reaction such as ionization or dissociation.

The annihilation mechanism requires an overlap of molecular orbitals, and hence a close approach by the annihilating pair of excited state molecules [787]. Excited state annihilation can occur if the excited state molecules diffuse together in solution, or if the excitations can themselves migrate within a crystal lattice or along a polymer chain.

Singlet–singlet (S–S) and triplet–triplet (T–T) annihilations are both homogenous annihilation processes if two photons are absorbed on the same chain (macromolecule). The singlet (triplet) excitation energy migrates as two independently excitons until they come close each other to undergo annihilation with formation of ground state S_0 (or T_0) and highly excited singlet (S^{**}) or triplet (T^{**}) state:

$$S_0 + S_0 \xrightarrow{+2h\nu} S^* + S^* \tag{3.60}$$

$$S^* + S^* \xrightarrow{\text{energy migration}} S^{**} + S_0 \quad \text{or} \tag{3.61}$$

$$S^* + S^* \xrightarrow{\text{ISC}} T^* + T^* \tag{3.62}$$

$$T^* + T^* \xrightarrow{\text{energy migration}} T^{**} + S_0 \ . \tag{3.63}$$

The highly excited S^{**} or T^{**} can be deactivated by emission of fluorescence or phosphorescence respectively, or dissociate into free radicals:

$$S^{**} \longrightarrow S_0 + \text{fluorescence} \tag{3.64}$$

$$T^{**} \longrightarrow S_0 + \text{phosphorescence} \tag{3.65}$$

$$S^{**} \ (\text{or } T^{**}) \longrightarrow \text{free radicals} \ . \tag{3.66}$$

The highly excited triplet state (T^{**}) can be deactivated with the formation of a ground state (S_0) and an excited singlet state (S^*) which then emits delayed fluorescence):

$$T^{**} \longrightarrow S^* + S_0 \tag{3.67}$$

$$S^* \longrightarrow S_0 + \text{delayed fluorescence} \ . \tag{3.68}$$

The singlet–singlet (S–S) annihilation is characterized by a strong non-linear intensity dependence of S^* fluorescence, transient absorption, photo-products formation and yield of intersystem crossing (ISC) to the triplet state.

The S–S and T–T annihilation processes have been found in several vinyl aromatic polymers [606]. Measurements of these processes are exceptionally difficult.

3.19 Energy Migration in Polymers

There are two types of energy migration in polymers [149, 245, 289, 290, 317]:

(i) intermolecular energy migration between donor and acceptor randomly distributed in an inert polymer matrix;
(ii) intramolecular energy migration (Fig. 3.10) from donor to acceptor along the chain (down-chain energy migration), and from donor to acceptor in a coiled chain (intracoil energy migration).

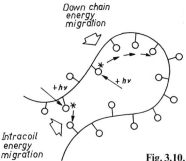

Fig. 3.10. Intramolecular energy migration in polymers

Intramolecular energy migration depends on:

(i)   chromophore orientation in different polymer conformations by a rotation of a chain segment (down-chain energy migration);
(ii) cross-chain segmental diffusion, which may occur within some "critical distance" between segments. The probability of segments lying within this "critical radius" of each other depends on coil density and hence molecular weight.

Energy absorbed by a donor is very often transferred not to acceptor but to the internal- or external-traps. There are the following types of traps:

(i)   impurities, which structurally do not belong to the polymer;
(ii) structural anomalies, e.g. branches, unsaturated bonds, catalyzers introduced in the polymerization process, groups which are formed during thermal processing and storage;
(iii) excimer forming sites.
(iv) charge-transfer (CT) sites.

Energy accumulated in a trap can be further:

(i)   transferred to a chromophore in a polymer or to another trap;
(ii) dissipated by non-radiative processes;
(iii) emitted as fluorescence and/or phosphorescence;
(iv) used for the dissociation of a trap into free radicals.

Several mechanism have been proposed for intramolecular energy migration in a polymer chain to a trap site (acceptor group).

(i)   Donor (D) is an external low molecular chromophore molecule which absorbs radiation and transfers its excitation energy (by electron exchange or resonance excitation) to a chromophoric group (M) which forms part of the molecular structure of the polymer. The excited chromophore (M*) transfers its excitation energy to the neighbouring chromophore and step by step energy migrates to a trap A (which is a final acceptor):

$$
\begin{aligned}
&h\nu \\
&D(S_0) \qquad\qquad D^*(S_1/T_1) \qquad\qquad D(S_0) \\
&-M-M-M-M-M-M-A \rightarrow -M-M- \rightarrow -M^*-M-M-M-A- \\
&-M-M-M^*-M-M-M-A \rightarrow -M-M-M-M^*-M-M-A- \\
&-M-M-M-M-M-M^*-M-A \rightarrow \\
&-M-M-M-M-M-M^*-A \rightarrow -M-M-A^*- \cdot
\end{aligned}
\qquad (3.69)
$$

(ii) Chromophore (M) which forms a part of the molecular structure of the polymer can directly absorb radiation and further transfer it by a migration process to a trap A:

hν

$$-M-M-M-M-M-M-A- \longrightarrow \qquad (3.70)$$

$$-M-\overset{*}{M}-M-M-M-M-A- \longrightarrow -M-M-\overset{*}{A}- \cdot$$

(iii) Successive energy migration from an excited chromophore (M) to an excimer forming site (M*M) and then to a trap A:

$$-M-M-M-(M^*M)-M-A- \longrightarrow$$

$$-M-\overset{*}{M}-M-M-M-M-A- \longrightarrow -M-M-A^*- \cdot \qquad (3.71)$$

(iv) Energy migration via excimer (M*M) which dissociate to M* which then transfers energy to a trap A:

$$-M-\overset{*}{M}-M-M-M-M-A- \longrightarrow$$

$$-M-M-M-(M^* M)-M-A- \longrightarrow -M-M-A^*- \cdot \qquad (3.72)$$

All chromophoric groups which separate an excited chromophore M* from an acceptor A (trap) are often called "spacers". The spacers can be photophysically active (participate in the downchain energy migration) or inactive (in such polymers only intracoil energy migration is possible).

Energy migration can be considered as "excitation hopping"

$$D^* + D + D \xrightarrow{k_M} D + D^* + D \xrightarrow{k_M} D + D + D^*. \qquad (3.73)$$

If energy migration does not result in re-excitation of a previously excited molecule then the number of hops per unit time is given by $k_M[D]$ and the number of hops per average donor (D) lifetime (τ) is given by $k_M[D]\tau$. The mean displacement ($\bar{r}$) of energy excitation during the average lifetime of a donor (τ) is then

$$\bar{r} = R\sqrt{k_M[D]\tau} \qquad (3.74)$$

where R is the distance the excitation travels per hop, [D] is the concentration of donor, and k_M is the rate constant of energy migration by excitation hopping.

For styrene-co-vinylbenzophenone, up to 10^3 jumps of the triplet energy migration from benzophenone to benzophenone is indicated.

Considering that $\bar{r}$ is the distance of net molecular energy migration, and $R_{0(DD)}$ is the critical transfer distance for which the probability of energy transfer (between D* and D) equals the probabilitty of deactivation of D* by all other processes, two extreme situations in energy migration may occur.

(i)   When $\bar{r} \ll R_0(DD)$, the D^* molecules are deactivated at distances of separation much large than the diffusional distances moved by the molecules or the excitation. In this case, the D^* and D molecules remain effectively stationary during the lifetime of D^*. These are conditions for which Förster [238], Perrin [223, 570], Dexter [200] or Inokuti and Hirayama [353] kinetics laws apply.

(ii) When $\bar{r} \gg R_{0(DD)}$, the excitation energy or D (or both) are effectively mobile and the energy migration rate (k_M) is not distance dependent. These are conditions for which the Stern-Volmer kinetic law may be applied.

The energy migration rate (k_M) between polymer-bound chromophores depends on [283]:

(i)   polymer-bound chromophore separation;
(ii) average mutual chromophore separations;
(iii) distribution of chromophore separations – a break in a sequence of chromophore can impose a barrier to energy migration;
(iv) spatial relationships that exist between adjacent or non-neighbouring chromophores.

The experimental results show that migration of energy prior to the transfer occurs in poly(vinylbenzophenone) and poly(phenyl-vinylketone) but not in poly(methylvinylketone), since the critical transfer distance determined in this latter case has the expected value for transfer by electron exchange interaction. The migration proceeds through the aromatic chromophores in such a way that the energy becomes delocalized in regularly ordered regions of the polymer. The behaviour of the polymer is thus intermediate between that of an amorphous medium and that of an ordered lattice. In the absence of aromatic groups, the migration of energy in polymers seems to be much less likely.

Energy migration in polymers depends on macromolecular structures:

(i)   type of polymer backbone;
(ii) type of bonding sequence by which the chromophores is attached to the backbone;
(iii) tacticity of the polymer;
(iv) molecular weight of the polymer, which affects both the coil density for a given solvent and the number of chromophores per coil increased molecular weight increases the probability of multiple photon absorption on the same coil;
(v)   coil density, which affects the extent of intracoil energy migration (cross-chain energy migration) between non-adjacent chromophores that are on different segments of the polymer chain – it is not possible to distinguish clearly intracoil energy migration from down-chain energy migration.

Energy migration in polymers in solution is affected in two ways. The first, excited state of chromophore groups, is quenched by:

(i)   external heavy atoms – iodo- or bromo-solvents and even CCl_4 quench effectively excited states – dichloromethane (CH_2Cl_2) does not quench excited states;

(ii) charge-transfer (CT) complex formation – solvents with cyano, amino or aromatic ester groups may give rise to an exciplex fluorescence or simply quench;

(iii) energy transfer to solvent molecule – in such a case it is difficult to excite the polymer-bound chromophore in such solvents.

The second is coil density. The thermodynamic quality of the solvent affects the coil density, and in turn affects the average chromophore separation [601]. The poorer solvents enhance the relative intensity of excimer fluorescence. Poor solvents increase efficiency of intracoil energy migration.

3.20 Energy Migration in Excimer-Forming Polymers

Vinyl aromatic polymers, where pendant chromophoric groups are bound to alternate backbone carbon atom show emission spectra which are composed from [167, 245, 246]:

(i) non-interacting chromophores (M*) (similar to their small molecular counterparts), and

(ii) broad structureless and somewhat red-shifted emission originating from the "excimers" (M*M), formed from the relatively weak binding between electronically excited chromophores (M*) (S_1) and their ground state counterparts M (S_0) [84, 85]:

$$M^* \ (^1S) \ + \ M(S_0) \ \underset{k_{MD}}{\overset{k_{DM}[M]}{\rightleftarrows}} \ (M^*M) \ (^1S) \ \xrightarrow{k_{ID}} \ 2\,M$$

$$k_{IM} \swarrow \quad \searrow k_{FM} \qquad \downarrow k_{FD}$$

$$M \qquad M + h\nu_M \qquad 2M + h\nu_D$$

(3.75)

(3.76)   (3.77)

where k_{DM} is the rate constant of excimer (M*M) formation, k_{MD} and k_{ID} are the rates of excimer deactivation, k_{IM} is the rate constant of the non-radiative deactivation of excited chromophore (M*) , and k_{FM} and k_{FD} are the rate constants of fluorescences from non-interacting chromophores (M*) and from the excimer (M*M) respectively.

An emission spectrum of poly(naphthyl methacrylate) consisting of monomer ("normal") and excimer ("broad") emissions is shown in Fig. 3.11.

Excimer fluorescence ($h\nu_D$) differs from the fluorescence from non-interacting chromophores ("normal fluorescence") ($h\nu_M$) because:

(i) it lies in the region of longer wavelengths than "normal fluorescence" because the excited state of the singlet excimer (M*M) (^1S) lies below the excited singlet state of chromophore M* (^1S);

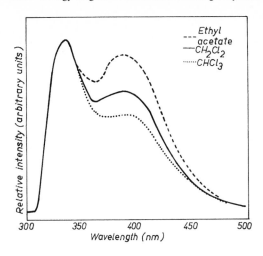

Fig. 3.11. Emission spectrum of poly(1–naphthyl methacrylate) at 23°C [314]

(ii) it shows a broad characteristic spectrum without vibrational structure because the ground state of the excimer is unstable;

iii) it depends on molecular structure – the number and nature of bond rotations necessary to create the excimer from the ground-state conformation, and the extent of excimer formation, are functions of structure;

(iv) formation of excimers in solution is diffusion-controlled and viscosity of solvent plays an important role;

(v)   it is dependent on temperature –

    (a) increasing temperature favours association, but at high temperatures thermal dissociation of excimers to excited and non-excited chromophores occurs,

    (b) if the temperature is so high that dissociation and association are rapid compared to the deactivation process, an equilibrium between excimer and excited chromophore is established, which leads to a common exponential decay of excited monomer and excited emission,

    (c) at low temperatures, when excimer dissociation is negligible, two exponential decays are observed – in the transition range the decay is non-exponential;

(vi) it is dependent on pressure which increases excimer formation in the dissociation equilibrium range;

(vii) excimer fluorescence may occur from various $(M^*\|M)$ and $(M^* \times M)$ dimer configurations, depending on the overlap conditions. The excimer fluorescence of aromatic polymers is due to intramolecular and intermolecular dimers suitably oriented and with sufficient freedom of motion to form excimers.

Absence of excimer fluorescence at room temperature does not imply the absence of excimer formation, because a small enthalpy of formation may lead to rapid dissociation of an excimer. If the lifetime of the excited chromophore

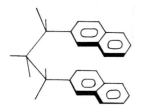

Fig. 3.12. Typical sandwich structure of the poly(2–vinyl naphtha-lene) excimer

is short, fairly high concentration is needed in order to observe the formation of excimer.

A critical requirement for the formation of excimers is the formation of a coplanar sandwich-like orientation of at least two aromatic groups (Fig. 3.12) affording the maximum π orbital overlap and in interchromophore separation in the range 3–7Å.

In polymers, the excimer formation is favoured by the trans-trans meso conformer [87]. This alignment of chromophores facilitates singlet energy migration along the polymer chain until the excitation is trapped at a chain conformation which is geometrically suitable for excimer formation [236, 246, 254]. Such a chain conformation is termed an "excimer-forming site". In vinyl aromatic polymers there are three different ways in which an excimer site can be formed (Fig. 3.13).

(i)   Intermolecular, by association between aromatic rings from two different polymer chains (intermolecular excimer). The number of these sites is directly proportional to the local concentration of aromatic rings.

(ii) Intramolecular (adjacent), by association between aromatic rings on ad-jacent repeating units in the chain backbone (intramolecular excimers). The number of these sites is independent of the concentration of chain molecules, and only dependent upon the steric effects associated with backbone bonds.

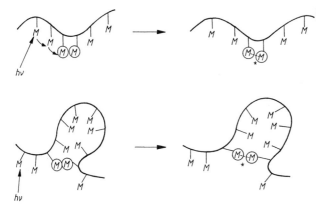

Fig. 3.13. Schematic description of excimer formation by energy migration between nearest neighbours along a polymer backbone and by intramolecular interactions between non-nearest neighbours

(iii) Intramolecular (non-adjacent), by association between aromatic rings on non-adjacent chain segments (intra-molecular excimers). The number of these sites is also independent on chain concentration. However, such an excimer site depends upon the polymer chain bending back upon itself, which is low probability for the rather stiff vinyl aromatic polymers.

In solid polymers a clear-cut distinction between inter- and intramolecular excimers is sometimes difficult.

Excimer formation is also controlled by the dynamic processes of macromolecules and the surrounding viscosity.

The intramolecular dynamics of macromolecules are controlled by the relaxation processes, which consist of:

(i) a molecular weight-dependent component which can be associated with the overall tumbling of the whole molecule;
(ii) a molecular weight-independent component which can be associated with local relaxation.

The observable relaxation time is considered to be the sum of these two separate mechanisms.

In fluid solutions of low viscosity, interconversion of chain conformations proceeds rapidly with the lifetime of a particular conformation limited by collision with solvent molecules [577]:

(i)   if the residence time of the random hopping singlet energy on a particular chromophore at a non-excimer forming site is long enough, solvent collisions with the polymer will cause a rotational sampling of chain conformations leading to possible excimer formation;
(ii) if the residence time is too short and/or rotational sampling rate too slow (such as occurs in high viscosity or at low temperature), exciplex sampling must occur by virtue of competitive trapping at a suitable excimer site formed prior to the arrival of the singlet energy – the reduced rotational freedom in the solid state precludes extensive rotational conformational sampling;
(iii) adjacent chromophores can also be in a marginal excimer-forming site which is not geometrically suitable to stabilize excimer formation appreciably.

Considering "exciton" (quantum of electronic energy) migration in solid polymers, there are two ways in which excimer can be formed [236, 249]:

(i)   an activated exciton migrates to preformed excimer-forming sites (temperature-dependent process) ("exciton migration model");
(ii) an unactivated exciton migrates to a marginal excimer site at which a geometrical barrier exists (temperature-independent process) ("site barrier model").

Excimer-forming sites are of greatest photophysical importance in vinyl aromatic polymers, even at low temperatures, because the electronic excitation energy in the form of excitons can migrate from chromophore to chromophore. This migration takes place non-radiatively by resonance energy transfer between an electronically excited chromophore and a ground-state chromophore. A theory for the incoherent transport and trapping of electronic excitations among chromophores attached to polymeric chains has been discussed in detail in several publications [110, 250–255].

The photophysical processes of polystyrene are dominated both in solution and in the solid state by inter- and intra- molecular excimer formation [322, 323, 571, 707]. The latter is associated with the occurrence of a specific conformation of the polymer chain in which two neighbouring phenyl groups share the excited state energy. Excimer formation has been investigated by a number of researchers interested in the configuration, distribution, dynamics and photodegradation of polystyrene and can be explained in terms of a combination of local rotational isomeric and longer range diffusional motions of the chain.

The kinetics of formation and decay of excimers in vinyl aromatic polymers can yield information on energy migration and segmental motion, although the interpretation can be difficult.

4 Photo-Oxidative Degradation

4.1 Introduction to Photochemical Reactions in Polymers

Photochemical reactions in polymers are those connected with changing chemical structure of the system.

A typical photochemical sequence can be divided into three stages:

(i) the absorptive act which produces an electronically excited state (cf. Chap. 1);
(ii) the primary photochemical process which involve electronically excited states (cf. Chap. 2);
(iii) the secondary or "dark" (thermal) reactions of the radicals, radical ions, ions and electrons produced by the primary photochemical process.

A direct photochemical process which causes polymer degradation is connected with bond dissociation. However, photodegradation is generally considered as a result of primary (photo-) and secondary (dark-) reactions which change a primary structure of a polymer by chain-scission, crosslinking and oxidative processes. These three processes can also differ in their localization. None of them, however, could be regarded as perfectly uniform throughout the solid specimen. First, both degradation and crosslinking take place predominantly in the amorphous part of the material. All radiation-induced processes start at specifically radiation sensitive sites (chromophoric groups) in the polymeric material. Therefore both chain scission and crosslinking are not introduced randomly. It is generally accepted that the photo-oxidation processes are initiated at the surface, which gives rise to gradient of deteriorated material across the specimen thickness (cf. Sect. 6.12).

4.2 Photochemical Laws

Four important photochemical laws can also be applied to polymer photochemistry:

(i) only the radiation absorbed by a molecule is effective in producing a photochemical change;
(ii) each photon or quantum absorbed activates one molecule in the primary excitation step of a photochemical sequence;

(iii) each photon or quantum absorbed by a molecule has a certain probability of populating either the lowest excited singlet state (S_1) or lowest triplet state (T_1);

(iv) the lowest excited singlet and triplet states are the starting points (in solution) of most organic photochemical processes.

4.3   Photochemical Activation

The selective nature of photochemical activation differentiates it from thermal activation. Absorption of a photon (quantum) of UV(visible) radiation can specifically excite (activate) a particular bond or group in a given macro-molecule. Thermal activation of the same macromolecule or a particular bond can only be achieved by an increase in the overall molecular energy of the system (environment).

Thermal activation energizes the polymer molecule through intermolecular vibrational energy transfer in collisions. Its total energy increases step by step as it ascends the vibrational ladder toward a dissociation threshold. As soon as the energy is just a little higher than the level of the excitation valley, the molecule may dissociate, but the products will have very little excess energy and will be produced in thermal equilibrium with the surrounding molecules.

In contrast, light absorption excites the macromolecule into upper vibra-tional levels in an excited electronic state in a single step. The absorption of radiation by a macromolecule occurs if the difference between two energy levels of absorbing bond (or group) in the macromolecule is equal to the energy of a photo (quantum):

$$\Delta E \; = \; h\nu \; = \; E_2 - E_1 \tag{4.1}$$

where h is Planck's constant, ν is the frequency (s^{-1}) at which absorption occurs, and E_2 and E_1 are the energies of a single absorbing species in the final and initial states.

If an excited macromolecule can find a dissociation path, it starts well up on the vibrational level, and it is almost invariably the case that more energy has been absorbed than the bare minimum necessary for dissociation. If the dis-sociation state does not occur with the ground state products, either or both of them may be electronically excited. Some of the remaining excess energy may be concentrated into vibration (or rotation) energy, while rest will appear as translational energy, divided between the products in inverse proportion to their masses. In general, the products of photodissociation are not produced in thermal equilibrium with the surroundings but are introduced into the system as "hot radicals", often chemically more reactive than their thermalized counterparts. The division of the excess energy between vibration, rotation and translation is controlled by the molecular dynamics of the dissociation process.

4.4 Photodissociation of a Molecule

Photochemical processes are chemical reactions in which electronically excited states resulting from absorption of irradiation are involved.

Excited singlet (S_1) and triplet (T_1) states can well be interpreted in the case of a diatomic molecules [62, 112, 770, 772]. Potential energy curves can be draw for each state S_0 (ground state), S_1 and T_1 of a diatomic molecules as shown in Figs. 4.1 and 4.2. Although the situation is not so simple for polyatomic molecules as for polymers, we can apply the above principle to these complex molecules at least as a general tendency. For polymers containing aromatic and unsaturated chromophoric groups, the energy necessary for excitation, dissociation, and ionization becomes larger in this order.

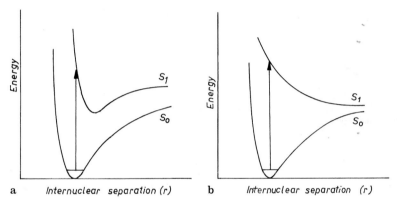

Fig. 4.1a, b. Photodissociation of a molecule through an excited singlet state (S_1): **a** by excitation above the dissociative limit of the upper state; **b** through a dissociative excited state [279]

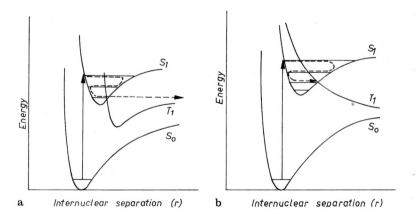

Fig. 4.2a, b. Photodissociation of a molecule through an excited triplet state (T_1): **a** by excitation above the dissociative limit of the upper state; **b** through a dissociative excited state [279]

Photodissociation of a molecule may occur both through singlet and triplet excited states. In Figs. 4.1 and 4.2 schematized potential energy curves of photodissociation of a molecule are shown. Figures 4.1a and 4.2a show that stable excited singlet or triplet states are reached but decomposition of the molecule occurs because absorption is to a point above the dissociation limit of the upper states. In this case energy may be dispersed as the kinetic energy of the fragments. Figures 4.1b and 4.2b show electronic excitation resulting from the formation of dissociative excited states such that the atoms repel each other at any separation distance (r).

Since excited triplet states have much longer lifetimes, more photo-dissociation processes are likely to occur from levels of that excited state.

Energies required for the direct dissociation of different bonds in organic molecules are shown in Table 1.1.

4.5   Quantum Yields of Photodegradation

Photodegradation processes can be followed by measuring different reactions which occur, e.g. chain scission, crosslinking, evolution of low molecular products (gases, liquids), etc. If it is possible to measure the extent of a given reaction and a number of radiation quanta absorbed, for each such reaction a quantum yield value can be evaluated [608, 609, 695]:
"quantum yield of chains scission"(ϕ_s) –

$$\phi_s = \frac{\text{number of macromolecules undergoing scission reaction}}{\text{number of quanta absorbed by the polymer system}} \; ; \quad (4.2)$$

"quantum yield of crosslinking"(ϕ_c) –

$$\phi_c = \frac{\text{number of macromolecules undergoing crosslinking reaction}}{\text{number of quanta absorbed by the polymer}}$$

$$(4.3)$$

"quantum yield of product (gaseous or liquid)"(ϕ_g or ϕ_l) –

$$\phi_g \text{ or } \phi_l = \frac{\text{number of molecules of lowmolecular product}}{\text{number of quanta absorbed by the polymer}} . \quad (4.4)$$

The extent of a given reaction occurring during the photodegradation of a given polymer at a specific wavelength can be measured by many different physical methods, whereas the number of quanta absorbed by the polymer system (in film or in solution) must be measured using calibrated photometer or chemical actinometers. Because it is extremely difficult to perform this type of absolute quantum measurements with a photometer routinely, indirect actinometer methods are usually preferred. They use as reference a reaction for which the quantum yield has previously been measured [602]. By determining the extent of chain scission (or product formation for known reaction) under conditions experimentally identical to those used for the unknown chemical reaction, it is possible to determine how many quanta of radiation have been used.

Studies of the variation in the quantum efficiencies of various primary photochemical processes in polymers show very little dependence on molecular weight [309, 319].

The knowledge of the value of quantum yield (ϕ) is important for the understanding of the mechanism and course of a photochemical reaction:

when $\phi = 1$, every absorbed quantum produces one photochemical reaction;

when $\phi < 1$, other reactions compete with the main photochemical reaction;

when $\phi > 1$, a chain reaction takes place.

The quantum yield defined above is wavelength-dependent. Most photochemical reactions in polymers are complex and occur in many stages; thus only in rare cases is the measured change that which was originally produced by the absorption of radiation.

The quantum yield of reaction should not be confused with the "chemical yields of reaction". A low quantum yield of reaction can sometimes lead to high chemical yield if no other reactions take place, and if one irradiation is long enough. However, the chemical yield is of little use in helping to establish the mechanisms of a photochemical reaction.

There are several potential complexities in the determination of quantum yields of photodegradation that arise when the progress of photoreaction is dependent on such factors as light filtering by products of photodegradation, secondary photoreactions, and inhomogenity of the polymer sample.

Number of quanta (photons) in a beam is commonly determined by using different types of actinometers [112, 602]. Most actinometric measurements are made under one of two limiting conditions:

(i) high absorbance (all the incident radiation is absorbed);
(ii) very low absorbance (the radiation absorption is directly proportional to concentration of absorbing chromophores):

$$I_a = I_0 - I_t = I_0(1 - 10^{-A})$$

$$= I_0(1 - \exp(-2.303\ A)) \approx 2.303\ I_0 A \ . \tag{4.5}$$

A number of expressions have been proposed for the change in rate of a photochemical reaction in homogenous media as the absorption characteristics of the medium alter with percentage conversion (typical case for polymer photodegradation) [107, 108].

4.6 Absorption of Radiation by Polymers

Most polymers contain only C–C, C–H, C–O, and C–Cl bonds and are not, therefore, expected to absorb radiation of wavelengths longer than 190 nm.

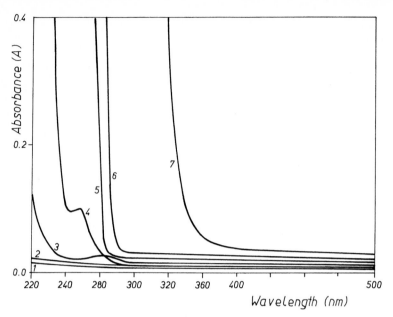

Fig. 4.3. Absorption spectra of different polymers: (*1*) polypropylene; (*2*) polyethylene; (*3*) poly(vinyl chloride); (*4*) poly(methyl methacrylate); (*5*) polystyrene; (*6*) polycarbonate and (*7*) polysulphone

The fact that polymers absorb radiation (Fig. 4.3) results from the presence of chromophores. Chromophores in polymers are of different types [30, 706].

(i) "Internal in-chain" and "end-chain" impurity chromophores (termed type A):

These isolated chromophores result from the polymerization process, products of thermal- and/or photo-oxidation (e.g. ketonic groups) or, occasionally, probes added deliberately.

(ii) "External" low molecular impurity chromophores:

X

_____

X

These chromophores are often present in the rest of polymerization catalyzers, commercial additives such as antioxidants, thermal- and photo-stabilizers, pigments, dyes, lubricants, plasticizers etc.

(iii) Chromophores which form part of the molecular structure of the polymer (termed type B):

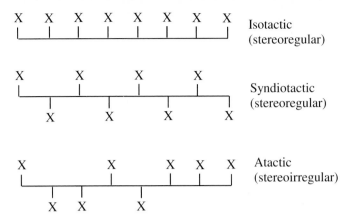

As a typical example of this type of polymer, poly(styrene) may be cited, in which pendant aromatic (chromophore) groups are stereochemically arranged along the backbone in a random (atactic), alternating (syndiotactic), or regular (isotactic) fashion. In this type of polymer the structural constraints of the polymer chain tend to confine the chromophore in spatial positions such that they can be expected to exhibit strong mutual interactions. These may depend strongly upon the relative orientation of the interacting chromophores, and the orientations themselves will usually be dependent upon the conformation of the polymer chain.

Absorption spectra of polymers in solutions differ sometimes because they can form charge-transfer complexes with solvent. Small but measurable differences in the absorption maxima are often observed when the spectrum of a given polymer is recorded in different solvents. The greatest differences are usually observed when comparing spectra in polar vs non-polar solvents, or protic vs non-protic solvents. However, the solubility of a polymer in a given solvent (polar or non-polar, protic or non-protic) depends on its structure.

The fact that polymers have absorption spectra covering an appreciable range of wavelengths means that electronic excitation can be produced by any radiation within this range.

The initiation of photodegradation of non-absorbing polymers of a given UV (light) radiation is commonly due to chromophores acting as photochemical impurities (present in polymers), which are capable of absorbing radiation energy, and undergoing photophysical processes and photochemical reactions.

4.7 Formation of UV/Light Absorbing Impurities

Internal and/or external impurities (very often containing UV/light absorbing chromophoric groups) are formed in several polymers (e.g. polyolefines, polystyrene, poly(vinyl chloride)) during polymerization processes and further

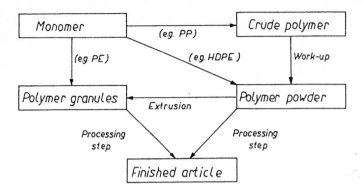

Fig. 4.4. Polyolefine polymerization processes: (*PE*) polyethylene; (*HDPE*) high density poly-ethylene and (*PP*) polypropylene [336]

processing and storage (Fig. 4.4). These impurities can be divided into two groups [599, 606, 608, 609, 638, 639].

(i)   "Internal impurities" which contain chromophoric groups, including
   (a)   hydroperoxides, carbonyl and unsaturated bonds,
   (b)   anomalous structural units such as branching (which originate in the polymerization process),
   (c)   catalyst residues attached to chain ends of macromolecules,
   (d)   charge-transfer complexes with oxygen (cf. Sect. 4.29).

(ii)   "External impurities" which may contain chromophoric and/or photo-reactive groups, including
   (a)   traces of catalysts, solvents, etc,
   (b)   additives (pigments, dyes, thermal stabilizers, antioxidants, photo-stabilizers, lubricants, plasticizers, etc,
   (c)   compounds from polluted urban atmosphere and smog, e.g. poly-nuclear hydrocarbons such as naphthalene, anthracene, etc. [624],
   (d)   traces of metals, metal oxides and metal salts from processing equipment and containers. The extrusion, milling, chopping and compounding steps involved in polymer processing can all introduce traces or even particles of such impurities into the polymer [609].

Free-radical polymerization of styrene is a commercially important process in which, since there is no rigorous exclusion of oxygen from the reaction, small amounts of oxygen dissolved in the styrene will be scavenged and incorporated into the polymer chain as in-chain peroxide linkages. Photolysis of these in-chain peroxides to give polymer alkoxy radicals, leading ultimately to hy-droperoxides, is a major route for initiation of the photo-oxidation of poly-styrene [430]. However, since most commercial polymerizations are carried out at elevated temperatures (as high as 200°C), the in-chain peroxides will undergo thermal fragmentation and disproportionation to form phenyl alkyl ketone end group chromophores [277].

Oxidation will occur if unstabilized polymer is exposed to gases containing even parts per million levels of oxygen for prolonged periods above 80°C, and over the course of months or years at ambient temperatures [336].

Oxidation also occurs during work-up (e.g. drying), granulation, silo storage and whilst awaiting conversion to the finished article during the service.

During processing, polymers are exposed to the following.

(i) High temperatures and oxygen, which cause thermal oxidation. The temperatures employed in the processing are generally 150–300°C depending on the polymer type, technology and the shape of the finished article. Sometimes very high temperatures, up to 300°C, are required for certain processing technologies, such as extrusion coating or melt spinning of fibres, or for injection moulding of parts with a very complicated geometry and shape. Polymer processing conditions are sometimes called the "processing history" of a polymer, which is very important in understanding photodegradation mechanisms.

(ii) The powerful shearing processes in a screw extruder. Shearing, processes cause stress at certain points in the chain, high enough to disrupt covalent bonds between atoms, which leads to bonds scission (so called "mechanodegradation").

4.8 Photoinitiation of Polymer Degradation

Photoinitiation of polymer degradation can be caused by:

(i) external low molecular impurities (RR') (cf. Sect. 4.30) which absorb UV and/or light radiation and produce low molecular weight radicals ($R\cdot$ and $R\cdot'$) which further react with a polymer (PH) producing a polymer alkyl radical ($P\cdot$) by the hydrogen atom abstraction reaction:

$$RR' \xrightarrow{h\nu} R\cdot + R\cdot' \tag{4.6}$$

$$PH + R\cdot \text{ (or } R\cdot') \longrightarrow P\cdot + RH \text{ (or } R'H) ; \tag{4.7}$$

(ii) internal in-chain and/or end-chain impurity chromophores present as part of a polymer structure (cf. Sect. 4.30) which absorb UV and/or light radiation and produce polymer alkyl radicals ($P\cdot$) and/or low-molecular radical fragments ($R\cdot$), e.g. methyl radical ($CH_3\cdot$):

$$\text{Polymer} \xrightarrow{h\nu} P\cdot + P\cdot \text{ (and/or } P\cdot + R\cdot) ; \tag{4.8}$$

(iii) direct dissociation of a given chemical bond, which has been excited to the excited singlet (S^*) or triplet (T^*) state:

$$\text{Polymer} \xrightarrow{h\nu} P\cdot + P\cdot ; \tag{4.9}$$

(iv) "charge-transfer complexes" formed between polymer and oxygen:

$$PH - O_2 \xrightarrow{h\nu} P\cdot + HO_2\cdot . \tag{4.10}$$

There has been much controversy in the literature about the relative importance of the various possible mechanisms for initiation (involving different impurities) but the controversy is practically irrelevant if initiation does not greatly influence the course, rate or extent of photo-oxidation [118, 119, 138, 606].

4.9   General Mechanism
of Polymer Photo-Oxidative Degradation

Photo-oxidative degradation of polymers, which includes such processes as photo-oxidation, chain scission, crosslinking and secondary reactions occurs by free radical mechanisms, similar in many respects to thermal oxidation.
The mechanistic steps of photo-oxidation includes:

$$\text{initiation} - \qquad PH \xrightarrow{h\nu} P \cdot \text{ (or } P \cdot + P \cdot) + (H \cdot) \qquad (4.11)$$

$$\text{chain propagation} - \quad P \cdot + O_2 \longrightarrow POO \cdot \qquad (4.12)$$

$$POO \cdot + PH \longrightarrow POOH + P \cdot \qquad (4.13)$$

$$\text{chain branching} - \quad POOH \longrightarrow PO \cdot + \cdot OH \qquad (4.14)$$

$$PH + \cdot OH \longrightarrow P \cdot + H_2O \qquad (4.15)$$

$$PO \cdot \longrightarrow \text{chain scission processes} \qquad (4.16)$$

$$\text{termination} - \qquad P \cdot + P \cdot \longrightarrow \qquad\qquad (4.17)$$

$$P \cdot + PO \cdot \longrightarrow \quad\text{crosslinking reactions} \qquad (4.18)$$

$$P \cdot + POO \cdot \longrightarrow \quad\text{to inactive products} \qquad (4.19)$$

$$POO \cdot + POO \cdot \longrightarrow \qquad\qquad (4.20)$$

where PH is the polymer, P· is the polymer alkyl radical, PO· is the polymer oxy radical (polymer alkoxy radical), POO· is the polymer peroxy radical (polymer alkylperoxy radical), POOH is the polymer hyderoperoxide, and HO· is the hydroxy radical.

Free radical nomenclature in this book is according to the IUPAC recommendations [651].

This mechanism applies to both thermal- and photo-oxidation of almost all polymers; however, the initiation steps differ in each case:

(i)   in thermal degradation, initiation results from the thermal dissociation of chemical bonds in macromolecules;

(ii) in photodegradation, initiation results from photodissociation of chemical bonds (from the singlet or triplet excited states) in a macromolecule.

Both thermal- and/or photo-degradation can also be initiated by the presence of external free radicals (R·) which are formed from the thermolysis and/or photolysis of some of impurities, additives or photoinitiators.

Theoretically, absorption of only one photon could cause the step by step degradation of a macromolecule but, in reality, the termination processes limit the degradation extent.

Photo-oxidation of polymers produces a complex mixture of different products. The identification of these products is the key point for the eva-luation of the mechanisms by which the photodegradation reactions occur. Two main difficulties exist that complicate the experimental identification of the products [443]:

(i)   the chemical changes that result from the photo-oxidation of polymers have to be studied on solid polymer matrix to take into account any perturbation that results from heterogenities of the solid state;
(ii) because mechanical failure of the photo-oxidized polymers occurs for very slightly oxidized samples, the reaction has to be limited to a small extent, which is usually less than 1%.

Photo-oxidative degradation generally causes main chain scission and cross-linking, the former being generally predominant in the presence of oxygen.

Many reviews [10, 48, 49, 96, 119, 126, 207, 273, 284, 285, 308, 309, 312, 315, 319, 333, 343, 405, 406, 442, 444, 452, 465–467, 484, 603, 611, 615, 636, 638, 639, 656, 668, 673, 692, 767] and books [15, 278, 302, 328, 363, 364, 369, 492, 508, 549, 565, 606, 608, 609, 632, 659, 675, 693, 784, 809] on polymer photodegradation have been published.

4.10   Chain Propagation

The key reaction in the "propagation sequences" is the formation of polymer peroxy (alkylperoxy) radicals (POO·) by reaction of polymer alkyl radicals (P·) with oxygen:

$$P· + O_2 \longrightarrow POO· \ . \tag{4.21}$$

This reaction is very fast, but diffusion controlled [198, 810, 811].

The next propagation step is the abstraction of a hydrogen atom by the polymer peroxy radical (POO·) to generate a new polymer alkyl radical (P·) and polymer hydroperoxide (POOH) [117, 272, 442]:

$$POO· + PH \longrightarrow P· + POOH \ . \tag{4.22}$$

Hydrogen atom abstraction occurs principally from the tertiary carbon atoms:

$$POO· + -CH_2 - \overset{\overset{\displaystyle R}{|}}{\underset{\underset{\displaystyle H}{|}}{C}} - CH_2 - \longrightarrow POOH + -CH_2 - \overset{\overset{\displaystyle R}{|}}{\underset{\displaystyle ·}{C}} - CH_2 - \ . \tag{4.23}$$

However, it may also occur from the secondary carbon atoms in methylene groups:

$$POO \cdot + -CH_2 - CH_2 \longrightarrow POOH + -CH_2 - \dot{C}H- \ . \tag{4.24}$$

Hydrogen atom abstraction reactions may also occur intramolecularly if a favourable stereochemical arrangement can be formed (Fig. 4.5).

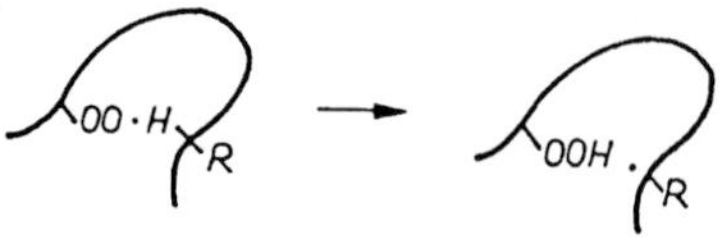

Fig. 4.5. Intramolecular hydrogen atom abstraction by the polymer peroxy radical (POO· .)

Polymer peroxy radicals can also participate in the termination reactions

$$POO \cdot + POO \cdot \rightarrow PO - OP + O_2 \tag{4.25}$$

$$POO \cdot + PO \cdot \rightarrow [POO - OP] \longrightarrow POP + O_2 \tag{4.26}$$

$$POO \cdot + P \cdot \rightarrow POOP \tag{4.27}$$

which compete with chain propagation.

4.11   Photochain-Oxidation Reaction

It has been suggested that photo-oxidative degradation of some polymers (e.g. polystyrene) may occur by so-called "photochain reactions" where a photon participates in each act of chain growth [280, 424]:

$$PH \xrightarrow{h\nu} P \cdot \xrightarrow{O_2} PO_2 \cdot \xrightarrow{h\nu, PH} P \cdot \ etc \ . \tag{4.28}$$

The presence of an effective photo-chain reaction means that photo-initiated oxidation is different from thermally-initiated oxidation, not only in the initiation stage but also in the propagation processes, as well as in the nature and yield of intermediate and final products. Chain reaction with a long kinetic chain leads to great polymer destruction, this being connected with the problem of destruction and stabilization of polymers and also with treatment of polymeric waste products.

4.12   Main Chain Scission

The "main chain scission" involves the breaking of the C–C bonds in the backbone. It occurs as a consequence of primary photophysical processes (direct photodissociation of a bond in the backbone) or as secondary processes (β-scission processes). It results in a decrease in the average molecular weight and can be represented by Fig. 4.6.

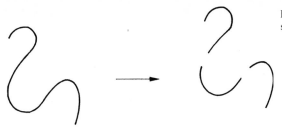

Fig. 4.6. A random main chain scission

Chain scission in polymeric alkoxy radicals (PO·) has generally been attributed to the β-scission reaction, which results in fragmentation of the polymer chain with formation of end carbonyl (or aldehyde) groups and end-polymer alkyl radicals:

$$-CH_2-\underset{\underset{O\cdot}{|}}{\overset{\overset{R}{|}}{C}}-CH_2-\overset{\overset{R}{|}}{CH}- \longrightarrow -CH_2-\underset{\underset{O}{||}}{\overset{\overset{R}{|}}{C}}+\cdot CH_2-\overset{\overset{R}{|}}{CH}- \qquad (4.29)$$

$$-CH_2-\underset{\underset{O\cdot}{|}}{CH}-CH_2-CH_2- \longrightarrow -CH_2-\underset{\underset{O}{||}}{\overset{\overset{H}{|}}{C}}+\cdot CH_2-CH_2-\;. \qquad (4.30)$$

It is generally agreed that decomposition of alkoxy radicals (β-scission) is the principal mode of chain scission in the photo-oxidative degradation of almost all polymers [18, 116, 118, 121, 122, 276, 606, 609, 632, 657, 792].

Alternative mode of chain scission involves a molecular decomposition of polymeric hydroperoxides (POOH) [281]:

$$-CH_2-\underset{\underset{R}{|}}{\overset{\overset{H}{\underset{|}{\overset{|}{O}}{\underset{|}{O}}}}{C}}-CH_2-\overset{\overset{R}{|}}{CH}- \longrightarrow -CH_2-\underset{\underset{R}{|}}{\overset{\overset{O}{||}}{C}}+ CH_2=\underset{\underset{R}{|}}{C}-\; + H_2O \qquad (4.31)$$

where $R = H$ or CH_3 .

4.13   Chain Branching

"Chain branching" by thermolysis and/or photolysis of polymer hydroperoxides (POOH) results in the formation of very reactive polymer oxy (alkyloxy) radicals (PO·) and hydroxy radicals (HO·) [118, 122, 276, 606, 609, 632, 792]:

$$POOH \xrightarrow{\Delta \text{ or } h\nu} PO\cdot + \cdot OH \ . \tag{4.32}$$

Polymer oxy radicals (PO·) and very mobile hydroxy radicals (HO·) abstract a hydrogen atom from the same or from a nearby polymer (PH) molecule:

$$PO\cdot + PH \longrightarrow POH + P\cdot \tag{4.33}$$

$$HO\cdot + PH \longrightarrow P\cdot + H_2O \ . \tag{4.34}$$

Polymer oxy radicals can also be a source of formation of in-chain ketone groups:

$$-CH_2-\overset{\overset{\displaystyle R}{\displaystyle |}}{\underset{\underset{\displaystyle O}{\displaystyle |}}{C}}-CH_2-\overset{\overset{\displaystyle R}{\displaystyle |}}{CH}-- \longrightarrow -CH_2-\overset{\overset{\displaystyle R}{\displaystyle |}}{\underset{\underset{\displaystyle O}{\displaystyle ||}}{C}}-CH_2-\overset{\overset{\displaystyle R}{\displaystyle |}}{CH}- +R\cdot \ . \tag{4.35}$$

However, the most important reaction of polymer oxy radicals (PO·) is the β-scission reaction (cf. Sect. 4.12).

4.14 Termination Reaction

The "termination" of the polymer radicals occurs by bimolecular recombination:

$$P\cdot + P\cdot \rightarrow PP \tag{4.36}$$

$$P\cdot + PO\cdot \rightarrow POP \tag{4.37}$$

$$P\cdot + POO\cdot \rightarrow POOP \tag{4.38}$$

$$PO\cdot + PO\cdot \rightarrow POOP \tag{4.39}$$

$$PO\cdot + POO\cdot \rightarrow POOOP \ (\text{or } POP + O_2) \tag{4.40}$$

$$POO\cdot + POO\cdot \rightarrow POO-OOP \ (\text{or } POOP + O_2) \ . \tag{4.41}$$

when the oxygen pressure is high (atmospheric pressure), the termination reaction almost exclusively occurs by reaction (4.41). At low oxygen pressure other terminations take place to some extent.

Several factors influence recombination reactions, such as:

(i)   cage effect;
(ii) effect of steric control;
(iii) mutual diffusion of reacting radicals – in solid polymers the recombination of polymer peroxy radicals (POO·) is controlled by the rate of their encounter with each other and is influenced by the intensity of molecular motion [219, 271];
(iv) structural parameters of the polymer matrix [270];
(v)   molecular-dynamical parameters of the polymer matrix.

If the polymer peroxy radicals are in the neighbouring positions, they can recombine to form stable cyclic peroxides (4.42) or epoxides (4.43):

$$-CH_2-\underset{\underset{\underset{\overset{\displaystyle\cdot}{O}}{|}}{\overset{\displaystyle O}{|}}}{\overset{\displaystyle R}{\underset{|}{C}}}-CH_2-\underset{\underset{\underset{\overset{\displaystyle\cdot}{O}}{|}}{\overset{\displaystyle O}{|}}}{\overset{\displaystyle R}{\underset{|}{C}}}-CH_2- \longrightarrow -CH_2-\overset{\displaystyle R}{\underset{\underset{O}{|}}{C}}-CH_2-\overset{\displaystyle R}{\underset{\underset{O}{|}}{C}}-CH_2- + O_2 \quad (4.42)$$

$$-CH_2-\underset{\underset{\underset{\overset{\displaystyle\cdot}{O}}{|}}{\overset{\displaystyle O}{|}}}{\overset{\displaystyle R}{\underset{|}{C}}}-\underset{\overset{\displaystyle\cdot}{\underset{|}{O}}}{\overset{\displaystyle R}{\underset{|}{CH}}}-CH- \longrightarrow -CH_2-\overset{\displaystyle R}{C}\underset{\diagdown\diagup}{\underset{O}{}}\overset{\displaystyle R}{CH}-CH- + O_2 . \quad (4.43)$$

Some of the termination reactions cause crosslinking, which creates a brittle polymer network.

4.15   Dark Processes in Photodegradation of Polymers

The fact that macroradicals in a solid polymer survive for a long time may account for oxidative degradation and crosslinking which continue for some time after irradiation, i.e. "dark processes" [384, 589]. However, some authors [79] question the possibility of chain scissions after irradiation. Consequently, the data seem to support the existence of a gradual recrystallization of radiation-scissioned chains over long ageing times [79, 567].

4.16   Hydrogen Atom Abstraction

Depending on the polymer chains structure, the hydrogen atom can be abstracted from the tertiary, secondary and even primary C–H sites.

(i)   The highest level of tertiary C–H sites in commercial and laboratory synthesized polyethylene samples is ~3/100 carbons. Several studies have shown that branched polyethylenes oxidize more rapidly then their linear analogues because of the presence of the tertiary C–H site at each branch point [193, 354].

(ii) Hydrogen atom abstraction by polymer alkoxy (PO·) and hydroxy (HO·) radicals from the photocleavage of hydroperoxy (POOH) groups in polypropylene occurs with a 50% probability at methyl groups. This means that up to 25% of the polymer peroxy radical (POO·) pairs produced in the initiation process may be pairs of primary peroxy radicals and the remaining 75% would be pairs between secondary, tertiary and mixed peroxy radicals. Due to the high radical concentration within such pairs of immediately adjacent radicals, the termination processes would be favoured over the slow hydrogen atom abstraction reaction of these radicals.

The intramolecular hydrogen atom abstraction by a polymer peroxy (POO·) radical can take place unless six-membered (or larger) rings are formed in the transition state. It is obvious that polymer segment conformation will determine the probability of formation of a transition complex of the optimum structure. Since structural relaxation is slow in a polymer matrix and the local mobility depends on the local conformation of a segment of macromolecule, the kinetics of intramolecular reactions will be influenced by the segment conformation of the macromolecule.

4.17 Formation of Hydroxy and/or Hydroperoxy Groups

Hydroxy (OH) and hydroperoxy (OOH) groups are formed in reactions between polymer oxy radicals (PO·) and polymer peroxy radicals (POO·) with the same and/or neighboring polymer molecule (PH), respectively:

$$PO\cdot \; + \; PH \longrightarrow POH \; + \; P\cdot \tag{4.44}$$

$$POO\cdot \; + \; PH \longrightarrow POOH \; + \; P\cdot \; . \tag{4.45}$$

Both groups can be formed along the polymer chain, or on its ends, but the latter is rather rare.

Hydroxy and hydroperoxy groups can also be formed in many reactions in which hydroxy (HO·) and/or hydroperoxy (HO$_2$·) radicals are involved (cf. Sect. 4.20).

The typical IR absorption bands for both groups are in the range 3400–3600 cm^{-1}. The most common method to differentiate these groups is to expose a sample to nitric oxide (NO) gas which reacts with OH and OOH groups giving nitrites and nitrates respectively [702]:

$$POH \; \xrightarrow[-20°C]{NO} \; PNO_2 \tag{4.46}$$

$$\text{(nitrite)}$$

$$POOH \xrightarrow[-20°C]{NO} PONO_2 \qquad\qquad (4.47)$$

(nitrate)

The sharp and intense IR spectra of these nitrates and nitrites can be used to differentiate between primary, secondary, and tertiary species [685, 748].

The secondary and tertiary hydroperoxide groups can also be detected by reaction with dimethyl sulfide (CH_3–S–CH_3) [230].

4.18   Homolytic Decomposition of Hydroperoxides

Hydroperoxides (POOH) undergo four types of homolytic reactions.

(i)   Unimolecular homolysis (photolysis and/or thermolysis) :

$$POOH \xrightarrow{\Delta,\ h\nu} PO\cdot + \cdot OH .\qquad\qquad (4.48)$$

In concentrated hydroperoxide solutions, second-order processes may also occur:

$$2\ POOH \longrightarrow PO\cdot + POO\cdot + H_2O .\qquad\qquad (4.49)$$

(ii)   Molecule-induced homolysis, when radicals are formed at high rate from the interaction of nonradical species:

$$POOH + A \longrightarrow PO\cdot + HOA\cdot .\qquad\qquad (4.50)$$

(iii)   Radical-induced decomposition [337, 791, 792]:

$$POOH + R'\cdot \longrightarrow PO_2\cdot + R'H \qquad\qquad (4.51)$$

$$2\ PO_2\cdot \longrightarrow 2\ PO\cdot + O_2 \qquad\qquad (4.52)$$

$$PO\cdot + POOH \longrightarrow POH + PO_2\cdot .\qquad\qquad (4.53)$$

(iv)   Free-radical displacement on O–O bond:

The structure of the hydroperoxide influences the course of its decomposition.

$$POOH + R'\cdot - \Big\langle\ \begin{array}{l} POR' + HO\cdot \qquad\qquad (4.54) \\[2em] R'OH + PO\cdot\ . \qquad\qquad (4.55) \end{array}$$

After oxygen–oxygen bond scission, several processes compete via radical–radical combination or disproportionation, radical abstraction from, or addition to available substrates, and alkoxyl (RO·) radical scission.

Decomposition of hydroperoxides in several polymers such as polyolefins [2, 124, 139–141, 172, 286, 336] and polystyrene [791, 792] have been extensively studied.

Intramolecular decomposition of secondary hydroperoxides leads to formation of ketone and vinyl groups [307]:

$$\begin{array}{c} H \\ \diagdown \diagup \\ C \\ \diagup \diagup \diagdown \diagdown \\ O \quad\quad CH_2 \\ | \quad\quad\quad | \\ O \quad\quad CH \\ \diagup \diagdown \diagup \\ H \quad H \end{array} \xrightarrow{h\nu} \begin{array}{c} O \\ || \\ -C-H \end{array} + CH_2{=}CH{-} + H_2O\,. \tag{4.56}$$

4.19 Photodecomposition of Hydroperoxide Groups

The light quanta produced by solar radiation (Fig. 1.1) are energetically sufficient to cleave PO–OH and also P–OOH, but hardly POO–H bonds, which have the dissociation energies 42–47 kcal mol^{-1} (PO–OH), 70 kcal mol^{-1} (P–OOH) and 90 kcal mol^{-1} (POO–H) [77, 78, 739]. The large difference in bond dissociation energy between PO–OH and P–OOH means that reaction with formation of PO· and ·OH radicals will predominate during light irradiation [428, 806].

Hydroperoxide groups are transparent at wavelengths > 340 nm (Fig. 4.7) and they have very low molar extinction coefficients (molar absorptivity) ($\varepsilon = 10 - 150$ l mol^{-1} cm^{-1}) at wavelength $\lambda = 340$ nm. The O–O bond has no low-lying stable excited state, and the potential energy surfaces of the first excited states are dissociative [225]. The quantum yield of the photocleavage hydroperoxide groups in the near-ultraviolet is close to 1.0 [121].

The photolysis of hydroperoxide groups under solar irradiation is a slow process. The average lifetime of an –OOH group in 10 μm polypropylene film under constant UV irradiation is ~ 25 h, equivalent to roughly 4–5 days in solar radiation [119, 124].

The most probable mechanism of photodecomposition of polystyrene hydroperoxides is via an energy transfer process (cf. Sect. 3.17). However, decomposition of hydroperoxides may also occur by the photoreduction mechanism in which a triplet state of a ketone group is involved [733]:

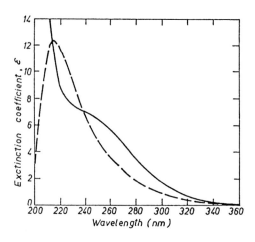

Fig. 4.7. Absorption spectra of (—) di-*tert*-butyl peroxide and (- - -) dimethyl peroxide

$$\text{P}-\text{OOH} + \overset{\diagdown}{\underset{\diagup}{\text{C}}} = \text{O}^* \longrightarrow \text{POO}\cdot + -\overset{\overset{\displaystyle\text{OH}}{\displaystyle|}}{\underset{\displaystyle|}{\text{C}}}\cdot \; . \tag{4.57}$$

The resulting tertiary peroxy radicals can interact to form a tetraoxide, which subsequently decomposes to yield a peroxide, polymer alkoxy radicals and oxygen [118]:

$$2\,\text{POO}\cdot \longrightarrow [\text{P} - \text{O} - \text{O} - \text{O} - \text{O} - \text{P}] \longrightarrow [\text{PO}\cdot + \text{O}_2 + \cdot\text{OP}]_{\text{cage}} \tag{4.58}$$

$$[\text{PO}\cdot + \text{O}_2 + \cdot\text{OP}]_{\text{cage}} \left\{ \begin{array}{ll} \text{P}-\text{O}-\text{O}-\text{P} + \text{O}_2 & \tag{4.59} \\[2em] 2\,\text{PO}\cdot + \text{O}_2 & \tag{4.60} \end{array} \right.$$

Polymer alkoxy radicals (PO·) may further abstract hydrogen from the same and/or neighbouring polymer molecules (PH) and produce a new polymer alky radical (P·) :

$$\text{PO}\cdot + \text{PH} \longrightarrow \text{P} - \text{OH} + \text{P}\cdot \; . \tag{4.61}$$

The large differences in behaviour between polymeric hydroperoxides in dilute solution and in thin film are related to increased competition from cross-linking reactions. Thus, while it would appear that the film is more photostable (self inhibition) than the solution, the effects of the degradation in the solid state are more noticeable and more serious, in terms of physical and mechanical properties (e.g. lower strength and increased embrittlement).

4.20   Reactions of Hydroxy (HO·) and Hydroperoxy (HO₂·) Radicals with Polymers

The main source of hydroxy (HO·) radicals is the unimolecular homolysis (photolysis and/or thermolysis) of polymeric hydroperoxides (POOH) (cf. Sect. 4.18).

When photo-oxidative degradation occurs in solution, hydroxy (HO·) radicals formed may escape out of the cage and avoid recombination with polymer alkyloxy (PO·) and/or other hydroxy radicals. In this process the fragments may never attain a separation of as much as one molecule diameter, and recombination takes place in a period that is longer than a vibration $(10^{-13}$ s) and less than the time between diffusive displacements $(10^{-11}$ s). When hydroxy (HO·) radicals undergo diffusive separation, i.e. escape from the cage, there is a high probability that they can react with a neighbouring polymer molecule (PH) and abstract a hydrogen atom [452]:

$$\text{PH} + \text{HO}\cdot \longrightarrow \text{P}\cdot + \text{H}_2\text{O} \; . \tag{4.62}$$

However, after a number of random displacements, hydroxy (HO·) radicals may encounter each other, giving hydroxy peroxide (H_2O_2):

$$HO· + HO· \longrightarrow H_2O_2 \ . \tag{4.63}$$

This cage effect is a function of the fluidity $(1/\eta)^{1/2}$ [103], where η is the viscosity coefficient.

Hydroxy (HO·) radicals can abstract hydrogen atoms from poly(ethylene oxide) [338, 382], and react with phenyl rings in aromatic polymers such as polystyrene [789] and poly(phenylene oxide) [383] causing ring opening reactions with the formation of α- and β-hydroxy mucondialdehydes (cf. Sect. 4.27)

It has been suggested that hydroperoxy (HO$_2$·) radicals can be formed in the photoinitiation step, according to the reaction [306, 452]

$$PH + O_2 \xrightarrow{h\nu} P· + HO_2· \ . \tag{4.64}$$

However, it is generally accepted that hydroperoxy (HO$_2$·) radicals are not capable of abstracting hydrogen atom from a polymer molecule.

Both hydroxy (HO·) and hydroperoxy (HO$_2$·) radicals can participate in the termination reaction with any available radical:

$$P· + HO· \longrightarrow POH \tag{4.65}$$

$$P· + HO_2· \longrightarrow POOH \tag{4.66}$$

$$PO· + HO· \longrightarrow POOH \tag{4.67}$$

$$PO· + HO_2· \longrightarrow PO - OOH \longrightarrow POH + O_2 \tag{4.68}$$

$$POO· + HO· \longrightarrow POO - OH \longrightarrow POH + O_2 \tag{4.69}$$

$$POO· + HO_2· \longrightarrow POO - OOH \longrightarrow POOH + O_2 \tag{4.70}$$

$$HO· + HO_2· \longrightarrow H_2O + O_2 \tag{4.71}$$

$$HO_2· + HO_2· \longrightarrow H_2O_2 + O_2 \ . \tag{4.72}$$

Hydroxy (HO·) radicals which are formed in the initial photolysis of polymeric hydroperoxides may also cause the decomposition of other POOH groups [792]:

$$POOH \xrightarrow{h\nu} PO· + ·OH \tag{4.73}$$

$$POOH + ·OH \longrightarrow POO· + H_2O \ . \tag{4.74}$$

Oxidative degradation caused by HO· and HO$_2$· radicals occurs with formation of aldehyde, carboxylic, hydroxyl and hydroperoxide groups, and with decreasing of the molecular weight and crosslinking reactions [382]. Polymer films exposed to HO· and HO$_2$· radicals (formed during photolysis of H_2O_2) shows formation of extended, deep surface cracks (Figure 4.8a, b) and a

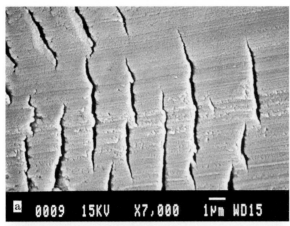

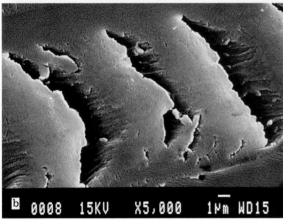

Fig. 4.8a, b. Scanning electron microscope photomicrographs of formation of extended, deep surface cracks in polymers photolysed in the presence of hydrogen peroxide: **a** poly(ethylene oxide) [382]; **b** polybutadiene (provided by Dr H. Kaczmarek, Copernicus University, Torun, Poland)

very rough surface (Fig. 4.9). In the case of polyisoprene, the oxidized amorphous surface has been etched, leaving crystalline spherulite structures (Fig. 4.10). Strongly oxidized polyacids like poly(acrylic acid) (Fig. 4.11a) and poly(methacrylic acid) (Fig. 4.11b) have a completely different surface morphology (cauliflower structures) in comparison to the flat surfaces of non-oxidized samples.

4.21 Formation of Carbonyl Groups

The carbonyl groups in the chain are formed mainly by the chain branching reaction (cf. Sect. 4.13), whereas end-carbonyl groups result from the β-scission process (cf. Sect. 4.12).

The "quantum yield for the formation of carbonyl (ketone and/or aldehyde) groups" (ϕ_{CO}) is given by

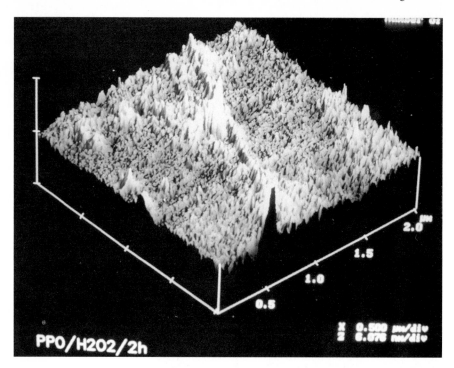

Fig. 4.9. Atomic force microscope photomicrograph of poly(phenylene oxide) photolysed in the presence of hydrogen peroxide [383] (provided by Dr H. Kaczmarek, Copernicus University, Torun, Poland)

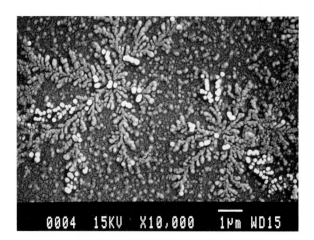

Fig. 4.10. Scanning electron microscope photomicrograph of the etched surface of polyisoprene photolysed in the presence of hydrogen peroxide (provided by Dr H. Kaczmarek, Copernicus University, Torun, Poland)

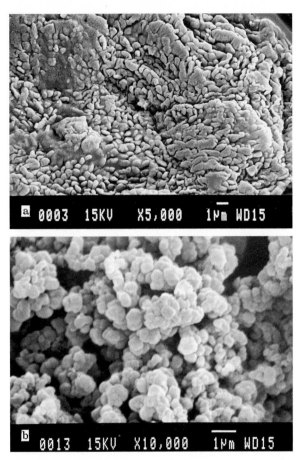

Fig. 4.11a, b. Scanning electron microscope photomicrographs: **a** poly(acrylic acid); **b** poly(-methacrylic acid) photolysed in the presence of hydrogen peroxide (provided by Dr H. Kacz-marek, Copernicus University, Torun, Poland)

$$\phi_{CO} = [CO]/E \tag{4.75}$$

where [CO] is the concentration of carbonyl groups in the polymer film (mol l^{-1}), and E is the energy of radiation absorbed (Einstein l^{-1}).

Identification of different carbonyl groups by IR spectroscopy has been given in detail elsewhere [608, 609]. However, there is an interesting qualitative reaction for the presence of aldehyde groups using the "Tollens Reagent" [566]:

$$P-C\overset{H}{\underset{O}{\diagdown}} + 2Ag(NH_3)_2OH \longrightarrow PCOO^- NH_4^+ + 2Ag + H_2O + 3NH_3 \; . \tag{4.76}$$

The polymer containing aldehyde groups is stained grey/black due to deposited Ag metal. This reaction occurs with aldehyde groups but not with ketone groups.

4.22   Formation of Carboxylic and Per-Carboxylic Groups

Polymer end-carboxylic groups can be formed by the following reactions [240, 799].

(i) Formation of polymer acyl radicals e.g. $\begin{matrix} O \\ \parallel \\ (P-C\cdot) \end{matrix}$ from the Norrish Type I reaction:

$$-CH_2-\overset{\overset{\displaystyle O}{\parallel}}{C}-CH_2- \longrightarrow -CH_2-\overset{\overset{\displaystyle O}{\parallel}}{C}\cdot +\cdot CH_2- \qquad (4.77)$$

or by hydrogen atom abstraction by radicals ($P\cdot, PO\cdot, POO\cdot$ or $R\cdot$) from the end-aldehyde group:

$$-CH_2-\overset{\overset{\displaystyle O}{\parallel}}{C}-H+P\cdot \text{ (or } PO\cdot, POO\cdot, R\cdot) \longrightarrow$$

$$-CH_2-\overset{\overset{\displaystyle O}{\parallel}}{C}\cdot +PH \text{ (or POH, POOH, RH) .} \qquad (4.78)$$

(ii) Polymer acyl radicals are further oxidized to polymer peracid radicals:

$$-CH_2-\overset{\overset{\displaystyle O}{\parallel}}{C}\cdot +O_2 \longrightarrow -CH_2-\overset{\overset{\displaystyle O}{\parallel}}{C}-OO\cdot \qquad (4.79)$$

which can further abstract hydrogen atom and form polymer peracids:

(iii)

$$-CH_2-\overset{\overset{\displaystyle O}{\parallel}}{C}-OO\cdot +PH \longrightarrow -CH_2-\overset{\overset{\displaystyle O}{\parallel}}{C}-OOH+P\cdot . \qquad (4.80)$$

Per-carboxylic groups undergo homolytic reaction giving polymer carboxy and hydroxy ($HO\cdot$) radicals:

$$-CH_2-\overset{\overset{\displaystyle O}{\parallel}}{C}-OOH \xrightarrow{\Delta, h\nu} -CH_2-\overset{\overset{\displaystyle O}{\parallel}}{C}-O\cdot +HO\cdot . \qquad (4.81)$$

(iv) Polymer end-carboxy radicals abstract hydrogen atoms and form polymer-end carboxylic groups:

$$-CH_2-\overset{\overset{\displaystyle O}{\parallel}}{C}-O\cdot +PH \longrightarrow -CH_2-\overset{\overset{\displaystyle O}{\parallel}}{C}-OH+P\cdot . \qquad (4.82)$$

(v) Polymer end-carboxy radicals behave as a source of polymer end-alkyl radicals and carbon dioxide:

$$-CH_2-\overset{\overset{\displaystyle O}{||}}{C}-O\cdot \longrightarrow -CH_2\cdot + CO_2 \cdot \tag{4.83}$$

(vi) In low-molecular peracids the well known mechanism can occur in which alkyl radical $(R\cdot)$ reacts with the peroxidic oxygen of the peracid in order to transfer the OH group and form an alcohol ROH:

$$R\cdot + \begin{array}{c} O \\ H \quad \searrow C-R \\ \diagdown \quad \diagup \\ O-C \end{array} \longrightarrow ROH + R-\overset{\overset{\displaystyle O}{||}}{C}-O\cdot \; . \tag{4.84}$$

Polymer peracid radicals can also be formed from the cage recombination of two polymer peroxy radicals $(POO\cdot)$ [232] :

$$[POO\cdot + \cdot OOP] \longrightarrow P-\overset{\overset{\displaystyle O}{||}}{C}-OO\cdot + HO_2\cdot \; . \tag{4.85}$$

Low efficiency of the carboxylic groups formation can be a result of a conversion of the polymer acyl radical $(-\overset{.}{C}=O)$ to polymer alkyl radical $(-\overset{.}{C}H-)$ under visible light irradiation [326, 327]:

$$-CH_2-CH_2-\overset{\overset{\displaystyle O}{||}}{C}\cdot \overset{h\nu}{\longrightarrow} -CH_2-\overset{.}{C}H-C\overset{\diagup\!\diagup O}{\underset{\diagdown H}{}} \; . \tag{4.86}$$

Polymer side-carboxyl groups can be formed by the following mechanism:

$$-CH_2-\overset{.}{C}H-CH_2- + CO \text{ (from the} \longrightarrow -CH_2-\overset{\overset{\displaystyle \overset{.}{C}O}{|}}{C}H-CH_2- \; . \tag{4.87}$$
$$\text{Norrish Type I}$$
$$\text{reaction)}$$

The presence of carboxylic groups can be confirmed by the reaction with NH_3 (formation of carboxylate ions) [652, 702, 799]:

$$R-\overset{\overset{\displaystyle O}{||}}{C}-OH \overset{NH_3}{\longrightarrow} R-\overset{\overset{\displaystyle O}{||}}{C}-O^- + NH_4^+ \tag{4.88}$$
$$(1550\text{-}1555 \text{ cm}^{-1})$$

and with SF_4 (formation of acyl fluorides) [115, 426, 702]:

$$R-\overset{\overset{\displaystyle O}{\|}}{C}-OH \xrightarrow{SF_4} R-\overset{\overset{\displaystyle O}{\|}}{C}-F \qquad . \qquad (4.89)$$
$$(1840\text{-}1850 \text{ cm}^{-1})$$

Dimethyl sulphide (CH_3–S–CH_3) can be used as a reagent to quantify peracids in the presence of hydroperoxides [821]:

$$-\overset{\overset{\displaystyle O}{\|}}{C}-OOH + CH_3-S-CH_3 \xrightarrow{fast} -\overset{\overset{\displaystyle O}{\|}}{C}-OH + CH_3-\overset{\overset{\displaystyle O}{\|}}{S}-CH_3 \qquad (4.90)$$

$$-\overset{\overset{\displaystyle |}{}}{\underset{\underset{\displaystyle |}{}}{C}}-OOH + CH_3-S-CH_3 \xrightarrow{slow} -\overset{\overset{\displaystyle |}{}}{\underset{\underset{\displaystyle |}{}}{C}}-OH + CH_3-\overset{\overset{\displaystyle O}{\|}}{S}-CH_3 \quad . \qquad (4.91)$$

Carboxyl groups in the presence of hydroxyl groups may also be differentiated by reaction with silver trifluoroacetate:

$$-COOH + CF_3COOAg \longrightarrow -COOAg + CF_3COOH \qquad (4.92)$$

or with trifluoroanhydride [201, 224, 536]:

$$-OH + (CF_3CO)_2O \longrightarrow -O-\overset{\overset{\displaystyle O}{\|}}{C}-CF_3 \qquad (4.93)$$

$$-COOH + (CF_3CO)_2O \longrightarrow -\overset{\overset{\displaystyle O}{\|}}{C}-O-\overset{\overset{\displaystyle O}{\|}}{C}-CF_3 \qquad (4.94)$$

$$-CH_2-\overset{\overset{\displaystyle \dot{C}O}{|}}{CH}-CH_2- \; + O_2 \longrightarrow -CH_2-\overset{\overset{\displaystyle OO\cdot}{\underset{\displaystyle CO}{|}}}{CH}-CH_2- \qquad (4.95)$$

$$-CH_2-\overset{\overset{\displaystyle OO\cdot}{\underset{\displaystyle CO}{|}}}{CH}-CH_2- \; + PH \longrightarrow -CH_2-\overset{\overset{\displaystyle OOH}{\underset{\displaystyle CO}{|}}}{CH}-CH_2- \; + P\cdot \qquad (4.96)$$

$$\begin{array}{cc} \overset{\displaystyle OOH}{\underset{\displaystyle |}{}} & \overset{\displaystyle \overset{\bullet}{O}}{\underset{\displaystyle |}{}} \\ CO & CO \\ | & | \\ -CH_2-CH-CH_2- \xrightarrow{\Delta, h\nu} -CH_2-CH-CH_2- + HO\cdot \end{array} \qquad (4.97)$$

$$\begin{array}{cc} \overset{\displaystyle \overset{\bullet}{O}}{\underset{\displaystyle |}{}} & \overset{\displaystyle OH}{\underset{\displaystyle |}{}} \\ CO & CO \\ | & | \\ -CH_2-CH-CH_2- + PH \longrightarrow -CH_2-CH-CH_2- + P\cdot \end{array} \qquad (4.98)$$

Polymer carboxylic groups can also be formed by the direct reaction of polymeric acyl radicals with hydroxy (HO·) radicals:

$$-CH_2-\overset{O}{\overset{||}{C}}\cdot + \cdot OH \longrightarrow -CH_2-\overset{O}{\overset{||}{C}}-OH \qquad (4.99)$$

$$\begin{array}{cc} \overset{\displaystyle \overset{\bullet}{C}O}{\underset{\displaystyle |}{}} & \overset{\displaystyle OH}{\underset{\displaystyle |}{}} \\ & CO \\ | & | \\ -CH_2-CH-CH_2- + \cdot OH \longrightarrow -CH_2-CH-CH_2- \end{array} \qquad (4.100)$$

The presence of carboxylic acid groups can also be indicative of chain scission processes, and the extent to which they are present generally correlates well with physical changes in the polymer [751].

4.23 Norrish Type I and Type II Reactions

Polymers containing ketone groups undergo two types of photochemical reactions:

(i) the Norrish type I (radical) reaction (also called α-cleavage), yielding two end polymeric radicals and carbon monoxide:

$$-CH_2-CH_2-\overset{O}{\overset{||}{C}}-CH_2-CH_2- \xrightarrow{h\nu} -CH_2-CH_2-\overset{O}{\overset{||}{C}}\cdot + \cdot CH_2-CH_2- \qquad (4.101)$$

$$-CH_2-CH_2-\overset{O}{\overset{||}{C}}\cdot \longrightarrow -CH_2-CH_2\cdot + CO \; ; \qquad (4.102)$$

(ii) the Norrish type II (non-radical) reaction which involves an intermolecular hydrogen abstraction by means of a cyclic six-membered intermediate, which subsequently rearranges to give a ketone and an olefin:

$$-CH_2-CH_2-\overset{\overset{\textstyle O}{\|}}{C}-CH_2-CH_2- \xrightarrow{h\nu}$$

$$-CH_2-CH_2-\overset{\overset{\textstyle O}{\|}}{C}-CH_3 + CH_2=CH- \ . \qquad (4.103)$$

The Norrish type I and type II reactions depend very much on polymer
structure.

In ethylene carbon monoxide copolymer the Norrish type I processes occur
with low probability because of the strong cage effect and reactivity of the
primary radicals produced. Ethylene carbon monoxide copolymer does not
undergo Norrish type II photolysis because of the absence of γ-hydrogens
[820]. In ethylene-vinyl ketone polymers, the photolysis of which yields one
polymeric and one small acetyl radical:

$$-CH_2-CH_2-\overset{\overset{\textstyle CH_3}{|}}{\underset{|}{\overset{|}{C=O}}}\overset{}{CH}-CH_2-CH_2- \xrightarrow{h\nu}$$

$$-CH_2-CH_2-\overset{\cdot}{CH}-CH_2-CH_2- + \cdot\overset{\overset{\textstyle CH_3}{|}}{C}=O \qquad (4.104)$$

which gives Norrish type I product with a very high efficiency because of
diffusion of the small acetyl radical away from its polymeric alkyl radical
counterpart.

The Norrish type II reaction depends also on the lifetime of the triplet
excited state of carbonyl group during which motions of macromolecule have
to permit the necessary six- (or seven-) membered intermediate:

(i) six-membered cyclic intermediate occurs in poly(phenyl vinyl ketone)
     [159];
(ii) seven-membered cyclic intermediate occurs in poly(methyl methacrylate-
     co-methyl vinyl ketone) [43].

Steady state (Stern-Volmer experiments, cf. Sect. 2.8) as well as laser spectroscopy on polymeric systems has revealed the various steps involved in the

Norrish type II photoscission in polymers [231]. Direct evidence that this photoelimination proceeds via a short-lived triplet state and a biradical has been found. The Norrish type II photoscission in polymers differs significantly from that occuring in small molecules. A process of energy transfer, partially diffusion controlled, is substantial in the case of small molecules. The process is different in a polymeric medium where the macromolecule viscosity and the migration of the excitation energy play prominent roles.

The Norrish type I and type II reactions are very limited below the glass transition temperature (T_g). At this temperature no molecular motion is possible. Above T_g quantum yields of the Norrish reactions are almost identical to those in solution at the same temperature (increases in mobility occurring at this transition) [159].

4.24 Formation of α, β-Unsaturated Carbonyl Groups

During photo-oxidative degradation of polymers, formation of α, β-unsaturated ketones and aldehydes has been observed:

$$-CH_2-CH=CH-\overset{\overset{\displaystyle O}{\|}}{C}- \qquad -CH_2-CH=CH-\overset{\overset{\displaystyle O}{\|}}{C}-H$$

$$(1685\ cm^{-1}) \qquad\qquad (1720,\ 1732\ cm^{-1})$$

the α, β-unsaturated carbonyl groups undergoing two possible photoreactions in the polymer [11, 13, 30]:

(i) the formation of β, γ-unsaturated carbonyls followed by Norrish type I and type II reactions:

$$-CH_2-CH=CH-\overset{\overset{\displaystyle O}{\|}}{C}-CH_2-\ \xrightarrow{h\nu}\ -CH=CH-CH_2-\overset{\overset{\displaystyle O}{\|}}{C}-CH_2-\quad(4.105)$$

$$\downarrow h\nu$$

$$\text{Norrish type I and II} \qquad (4.106)$$
$$\text{reactions;}$$

(ii) crosslinking between adjacent α, β-unsaturated carbonyls to produce saturated carbonyls:

$$\begin{array}{l} -CH_2-CH=CH-\overset{\overset{\displaystyle O}{\|}}{C}-CH_2- \\[4pt] -CH_2-CH=CH-\underset{\underset{\displaystyle O}{\|}}{C}-CH_2- \end{array} \xrightarrow{h\nu} \begin{array}{l} -CH_2-CH-CH-\overset{\overset{\displaystyle O}{\|}}{C}-CH_2- \\[2pt] \qquad\quad |\ \ \ | \\[2pt] -CH_2-CH-CH-\underset{\underset{\displaystyle O}{\|}}{C}-CH_2-\ \cdot \end{array} \qquad (4.107)$$

4.25 Conversion of Polymeric Free Radicals Under UV/Visible Radiation and Warming

Electron spin resonance spectroscopy shows that polymeric allylic radical $(-CH = CH - \dot{C}H-)$ can be converted to polymer alkyl radical $(-CH_2 - \dot{C}H - CH_2-)$ under UV irradiation. Further warming reconverts the polymer alkyl radicals to stable allylic radicals [688, 689]:

$$-CH_2-CH=CH-\dot{C}H-CH_2-CH_2- \underset{\text{heat } (\Delta)}{\overset{h\nu}{\rightleftharpoons}}$$
(septet-line ESR spectrum)

$$-CH_2-CH=CH-CH_2-\dot{C}H-CH_2- \quad .$$
(sextet-line ESR spectrum)   (4.108)

Irradiation with visible light (> 390 nm) causes chain scission of the polymer allylic radical with formation of polymer end-alkyl radical $(CH_3 - \dot{C}H - CH_2-)$:

$$-CH_2-CH=CH-\dot{C}H-CH_2-CH_2-CH_2- \overset{h\nu}{\rightarrow}$$
(septet-line ESR spectrum)

$$-CH_2-CH=CH-CH=CH_2 + CH_3-\dot{C}H-CH_2- \quad . \quad (4.109)$$
(octet-line ESR spectrum)

After short warming to 273 K the polymer end-alkyl radical is converted to another type of polymer chain-alkyl radical $(-CH_2 - \dot{C}H - CH_2-)$:

$$CH_3-\dot{C}H-CH_2-CH_2- \overset{\Delta}{\rightarrow} CH_3-CH_2-\dot{C}H-CH_2- \quad .$$
(octet-line ESR spectrum)   (sextet-line ESR spectrum)   (4.110)

Prolonged warming to 273 K of the polymer end-alkyl radical causes formation of polymeric allylic radical:

$$CH_3-\dot{C}H-CH_2-CH_2-CH_2-CH=CH-CH_2- \overset{\Delta}{\rightarrow}$$
(octet-line ESR spectrum)

$$CH_3-CH_2-CH_2-CH_2-\dot{C}H-CH=CH-CH_2- \quad .$$
(septet-line ESR spectrum)   (4.111)

The activation energy of these processes has been found to be 18 kcal mol^{-1} [690].

4.26   Photodegradation of Chlorinated Polymers

Dehydrochlorination is the most characteristic reaction observed during UV irradiation of chlorinated polymers, e.g. poly(vinyl chloride). This reaction leads to the formation of polyene structures $(-CH = CH-)_n$ with n ranging from 2 to 13, which are responsible for yellow-red coloring of a polymer:

$$-CH_2-CH-CH_2-CH- \xrightarrow{h\nu} -CH_2-CH-\dot{C}H_2-CH- + Cl\cdot$$
$$\qquad\quad | \qquad\qquad | \qquad\qquad\qquad\qquad\qquad\quad |$$
$$\qquad\quad Cl \qquad\qquad Cl \qquad\qquad\qquad\qquad\qquad\quad Cl$$

$$-CH_2-\dot{C}H-CH_2-CH- + Cl\cdot \longrightarrow -CH_2-CH=CH-CH- + HCl \xrightarrow{h\nu}$$
$$\qquad\qquad\qquad\qquad | \qquad\qquad\qquad\qquad\qquad\qquad\qquad\qquad |$$
$$\qquad\qquad\qquad\qquad Cl \qquad\qquad\qquad\qquad\qquad\qquad\qquad\qquad Cl$$

$$(-CH=CH-CH=CH-CH=CH-)_n + nHCl \quad . \tag{4.112}$$

However, when this process is carried out in the presence of air, it leads to the typical photo-oxidative reactions with formation of carbonyl, carboxylic acid, hydroxyl and hydroperoxide groups [178–180, 187, 190–192, 259, 263–267, 610, 611, 627].

Photo-dehydrochlorination of chlorinated poly(vinyl chloride) leads to the formation of chlorinated polyene sequences [186, 192]:

$$-CH-CH- \xrightarrow{h\nu} -(C=CH)_n- + nHCl \quad .$$
$$\quad | \quad\quad | \qquad\qquad\qquad | \tag{4.113}$$
$$\quad Cl \quad\;\; Cl \qquad\qquad\quad Cl$$

The dehydrochlorination chain reaction is ten times more efficient in chlorinated poly(vinyl chloride) than in non-chlorinated polymer.

Samples which contain chlorinated polyenes can be further completely dehydrochlorinated by laser radiation to graphitized structures [181–183, 185, 188, 189, 691]:

$$-(C=CH)_n- \xrightarrow{h\nu} =(C=C)_n= + nHCl \quad . \tag{4.114}$$
$$\quad | $$
$$\quad Cl$$

4.27   Phenyl Ring-Opening Photo-Reactions

Phenyl rings (chromophores which form part of the molecular structure) in some polymers like polystyrene [616, 789, 790], poly(phenylene oxide) [652] and many others can be photo-oxidized by the ring-opening reaction, analogical to that observed during photo-oxidation of benzene [788]. The ring-opening reactions of the phenyl group results in the formation of conjugated dialdehydes (α-, and β-mucondialdehydes) :

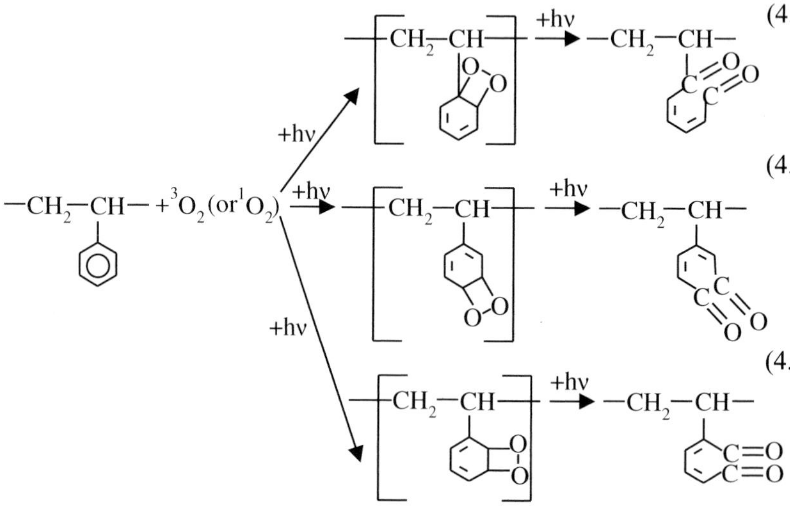

(4.115)

(4.116)

(4.117)

The IR spectroscopy of chain deuterated polystyrene [458], ESCA spectra [145, 146, 568, 797], measurements of interfacial tension and surface hydrophilicity [735], and studies on photo-oxidation of low-molecular model compounds [459, 460] strongly support the ring-opening in polystyrene. Formation of mucondialdehyde chromophores can be responsible for the yellowing of photo-oxidized polystyrene [616] and poly(phenylene oxide) [383].

4.28   PhotoFries Rearrangement in Polymers

The photoFries rearrangement (common in organic photochemistry [74]) also occurs in the case of some polymers with suitable molecular structures like polycarbonates [3, 75, 142, 226–228, 349, 350, 444, 562, 584, 653–655, 682, 757, 777], poly (4,4'-diphenylpropane isophthalate) [472], poly(fluorenone isophthalate) [456], poly(1, 3-phenylene isophthalamide) [120], polyurethanes [68, 260–262, 346, 347, 578] and epoxy resins [69–72].

The photoFries reaction (in the case of polycarbonates irradiated with radiation > 240 nm) occurs by the following mechanism [653–655]:

$$(4.118)$$

The photoFries rearrangement apparently involves the cleavage of a δ bond in the singlet excited state as an efficient process [1, 385]. The singlet state, according to available evidences [74] may be either n, π^* or π, π^* configuration.

Bond scission occurs primarily at the aromatic ether oxygen carbon bond, and causes degradation of a polymer backbone [324, 350, 653–655, 758].

By these two processes (photoFries rearrangement and degradation), are formed phenylsalicylate, dihydroxybenzophenone, mono-or dihydroxy-biphenyl, hydoxydiphenyl ether groups, phenolic end groups, CO and CO_2.

The photoFries reaction in solid polymer matrices indicates that there is very little effect due to the glass transition temperature (T_g). The photoFries reaction proceeds by means of a "caged radical" mechanism. (cf. Sect. 5.8).

In the case of poly(phenyl acrylate), only a small rotation of the phenoxy radical intermediate is necessary to provide for the rearranged product yielding an acetophenone from the original phenyl ester.

4.29   Photo-Oxygenation of Polymers by Singlet Oxygen

Oxygen, in its singlet states $^1O_2 \, (^1\Delta_g)$ and $^1O_2(^1\Sigma_g^+)$ is a versatile reagent with polymers containing allylic and/or diene unsaturated bonds [125, 297, 387, 464, 470, 594, 600, 604, 613, 618–623, 625, 626, 630, 632, 633, 635, 671, 832].

Depending on the availability of an allylic hydrogen, substituent polymers react with it to give either dioxetanes or hydroperoxides, whereas polydienes yield endo-peroxides [298, 386, 511, 513, 514, 622, 744–747]:

$$(4.119)$$

$$(4.120)$$

Originally, the ene-type reaction of singlet oxygen to produce polymeric hydroperoxide was regarded as being a concerted process. However, evidence now exists that an ionic or biradical intermediate is involved. Hydroperoxidation is an electrophilic reaction, the rate of which is governed by the energy of the highest occupied molecular orbital (HOMO) of the double bond [812].

Singlet oxygen ($^1O_2$) in polymers can be formed by energy transfer mechanism from impurities or specially added photosensitizers (e.g. dyes). The following mechanisms for the sensitizer (S) are allowed:

$$S_0 \xrightarrow{h\nu} {}^1S\,(S_1) \xrightarrow{ISC} {}^3S\,(T_1) \tag{4.121}$$

and, where ISC is the intersystem cross:

$${}^1S + O_2 \longrightarrow {}^3S + {}^1O_2\,({}^1\Delta_g, {}^1\Sigma_g^+) \tag{4.122}$$

$${}^1S + O_2 \longrightarrow S_0 + {}^1O_2({}^1\Delta_g, {}^1\Sigma_g^+) \quad \text{and} \tag{4.123}$$

$${}^3S + O_2 \longrightarrow S_0 + {}^1O_2({}^1\Delta_g, {}^1\Sigma_g^+) \;. \tag{4.124}$$

These energy transfer reactions are enhanced if, between excited sensitizer molecule and oxygen, there is formed an intermediate contact complex or a charge-transfer (CT) complex, with their further dissociation in the excited state [392–395, 397, 766]:

$$S_0 + O_2 \longrightarrow (S_0 \cdots O_2)\,(S_0) \xrightarrow{h\nu} (S \cdots O_2)^*\,(S_1 \text{ or } T_1)\;. \tag{4.125}$$
$$\text{CT complex}$$

Molecular oxygen can form charge-transfer complexes with a number of solid polymers, such as polyethylene [306, 768], polypropylene [17], poly (4-methyl-1-pentene) [807] and polystyrene [6, 468, 531, 532, 543, 616, 628, 629, 813].

Theoretical studies show that the CT state, which can be represented as an excited-state complex with radical cation and radical anion character $(M^{\cdot +} O_2^{\cdot -})$, will interact, often substantially, with other excited electronic states of the oxygen-organic (polymeric) molecule (M) complex [390, 391, 398, 544, 766]. In certain polar solvents, the CT complex is in equilibrium with the solvated organic radical cation $(M^{\cdot +})$ and the superoxide ion $(O_2^{\cdot -})$ [486]. The photolysis of the CT band results in the formation of two independent and reactive forms of molecular oxygen, i.e. singlet oxygen $^1O_2(^1\Delta_g)$ and oxygen radical anion $(O_2^{\cdot -})$. It has been proved that irradiation into the polymer-oxygen charge transfer (CT) absorption band of polystyrene creates singlet oxygen $^1O_2(^1\Delta_g)$ [543].

Theoretical investigations based on quantum mechanics [392, 393, 397] and experimental studies [394] indicate that, in the case of sensitizers having high energy in the triplet state $(E_T > 40 \text{ kcal mol}^{-1})$, singlet oxygen $^1O_2(^1\Sigma_g^+)$ is the main product in the energy transfer reaction. The amount of singlet oxygen $^1O_2(^1\Delta_g)$ formed in this reaction is ten times lower. In the case sensitizers with triplet energy (E_T) between 22 and 38 kcal mol^{-1}, only singlet oxygen $^1O_2(^1\Delta_g)$ is formed. Both forms of singlet oxygen are formed when the triplet energy of a sensitizer E_T exceeds 38 kcal mol^{-1}. These results indicate that the form of singlet oxygen is dependent on the excitation energy of the sensitizer (Fig. 4.12).

Triplet energy transfer from an excited dye to molecular oxygen (reaction 4.124) can be a very efficient process [148, 216, 356, 357, 604, 678, 679]. The quantum yield of $^1O_2(^1\Delta_g)$ depends on dye sensitizer (the decay rate of the dye triplet state) and on the solvent, and can range from 0 to 1 [414]. However, the quantum yield of $^1O_2(^1\Delta_g)$ in the singlet energy transfer reaction (reaction 4.123) can be as large as 2.0 [202, 414].

The very rapid deactivation of singlet oxygen $^1O_2(^1\Sigma_g^+)$ leads to the formation of singlet oxygen $^1O_2(^1\Delta_g)$:

$$^1O_2(^1\Sigma_g^+) \longrightarrow \; ^1O_2(^1\Delta_g) \; . \tag{4.126}$$

In the gas phase (under low pressure) singlet oxygen $^1O_2(^1\Delta_g)$ is extremely long-lived (around 45 min). The singlet oxygen $^1O_2(^1\Sigma_g^+)$ has a shorter lifetime

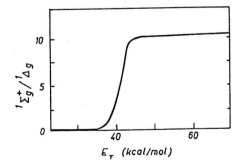

Fig. 4.12. Predicted variation in the $^1O_2(^1\Sigma_g) + \, ^1O_2(^1\Delta_g)$ ratio with the sensitizer triplet excitation energy (E_T) [395]

Solvent or solvent mixture	τ (s)	**Table 4.1.** Relative lifetimes of $^1O_2$ in different solvents
H$_2$O	2.0×10^{-6}	
H$_2$O + CH$_3$OH (1 : 1)	3.5×10^{-6}	
D$_2$O	2.0×10^{-5}	
CH$_3$OH	5.0×10^{-6}	
CH$_3$CH$_2$OH	5.6×10^{-6}	
Benzene	1.25×10^{-5}	
Cyclohexane	1.7×10^{-5}	
Toluene	2.0×10^{-5}	
Iso–octane	2.0×10^{-5}	
Dioxane	3.2×10^{-5}	
CHCl$_3$	6.0×10^{-5}	
CHCl$_3$ + CH$_3$OH (9 : 1)	2.6×10^{-5}	
Chlorobenzene	9.1×10^{-5}	
CH$_2$Cl$_2$	1.05×10^{-4}	
CS$_2$	2.0×10^{-4}	
CCl$_4$	7.0×10^{-4}	
CF$_3$Cl (Freon11)	1.0×10^{-3}	
CCl$_2$FCClF$_2$ (Freon113)	2.2×10^{-3}	

(around 7 s) and is rapidly quenched by water vapour. The lifetime of singlet oxygen in solution is strongly affected by the nature of the solvent (Table 4.1).

In solid polymers, both the rate of singlet oxygen $^1O_2(^1\Delta_g)$ formation and the yield of singlet oxygen in an energy transfer reaction will depend on [542]:

(i)   the diffusion coefficient for oxygen which is determined by the sample rigidity, the degree of polymerization (molecular weight), and the presence of plasticizers;
(ii) the concentration of oxygen in a sample;
(iii) the sample temperature.

The rate of singlet oxygen $^1O_2(^1\Delta_g)$ formation in a polymer matrix is dependent on:

(i)   the atomic-molecular composition of a polymer matrix;
(ii) the sample temperature;
(iii) the presence of quenchers (internal and/or external impurities).

The intrinsic lifetimes of singlet oxygen $^1O_2(^1\Delta_g)(k^{-1}_{decay})$ in different polymers are shown in Table 4.2.

In polymers, light absorbing impurities (external and/or internal (structure irregularities)) can act as potential photosensitizers for generation of singlet oxygen. Some examples of impurities and anomalous macromolecular structures that have been suggested include:

Polymer	Intrinsic lifetime (τ) (μ s)
Poly(4-methyl-1-pentene)	18 ± 2
Poly(methyl methacrylate)	22 ± 3
Poly(ethyl acrylate)	31 ± 1
Polystyrene	19 ± 2
Poly(deuteriopolystyrene)	250 ± 15
Poly(α-methylstyrene)	18 ± 2
Polycarbonate	29 ± 2
DuPont Teflon AF Type 1600	1700 ± 100

Table 4.2.   Singlet oxygen $^1O_2$ (^1D_g) lifetimes in solid polymers [542]

$$\left[CF_2CF_2 \right]_n \quad \left[\begin{array}{cc} CF & - CF \\ | & | \\ O & O \end{array} \right]_m$$

$$F_3C \quad CF_3$$

Polyurethane	16 ± 2
Poly(dimethyl siloxane)	46 ± 1

(i)   carbonyl groups on chain backbones whose origins can be attributed to the presence of oxygen or carbon monoxide in polymerization processes [387, 765];

(ii)   oxidation products (carbonyl groups, aromatic hydrocarbons) that can arise during processing of the material [387];

(iii)   aromatic hydrocarbons that might originate from the atmosphere (due to burning of hydrocarbon fuels or from car exhausts) [387, 624].

The singlet oxygen ($^1O_2$) formation in solid polystyrene can proceed by two independent pathways; photosensitization (by impurities) and CT state dissociation, and the relative importance of these pathways depends on the polymer sample history and method of preparation [543].

4.30   Photoinitiated Degradation of Polymers

Photodegradation (photo-oxidative degradation) of polymers can be photochemically initiated by free radicals formed from a photoinitiator added to a polymer sample [596–600, 606, 609, 612, 632, 636].

The photoinitiation of free radical abstraction of a hydrogen atom from a polymer molecule (PH) occurs in two steps.

(i)   Formation of free radicals (R·) from the photoinitiator (I):

$$I \xrightarrow{h\nu} R_1\cdot + R_2\cdot . \tag{4.127}$$

The rate of initiator decomposition (r_d) is given by

$$r_d = \phi_d I_a \tag{4.128}$$

where ϕ_d is the quantum yield of photodecomposition of initiator, and I_a is the intensity of UV (light) radiation absorbed by a photoinitiator.

(ii) Abstraction of a hydrogen atom from a polymer molecule (PH):

$$R_1 \cdot \text{ (or } R_2 \cdot) + PH \longrightarrow R_1 H \text{ (or } R_2 H) + P \cdot \ . \tag{4.129}$$

The rate of the polymer alkyl radical (P·) formation is given by

$$r_i = k_i [R \cdot] \tag{4.130}$$

where k_i is the polymer alkyl radical formation constant ($1 \ \text{mol}^{-1} \text{s}^{-1}$).

If both radicals $R_1 \cdot$ and $R_2 \cdot$ are able to abstract hydrogen with the same efficiency then the effective rate of polymer alkyl radical formation is

$$r_i = 2 \, k_i [R \cdot] \ . \tag{4.131}$$

In practice, the kinetics of these two steps are more complicated because:

(i)   radicals of different types, formed from photocleavage of the photo-initiator, have different reactivities with polymer molecules in the hydrogen abstraction (from secondary and tertiary carbon atoms;

(ii)   the recombination of $R_1 \cdot$ and $R_2 \cdot$ radicals may occur ("cage effect") (cf. Sect. 5.8);

(iii) free radicals $R_1 \cdot$ and $R_2 \cdot$ can be terminated by different polymer radicals, such as P·, PO·, or POO·;

(iv)   radicals $R_1 \cdot$ and $R_2 \cdot$ can react with oxygen, giving peroxy radicals $R_1 OO \cdot$ and $R_2 OO \cdot$;

(v)   radicals $R_1 \cdot$ and/or $R_2 \cdot$ can be scavenged by additives (antioxidants, thermostabilizers, photostabilizers, inhibitors) present in a polymer sample (plastic).

Radicals can be formed from the singlet (S_1) and/or triplet (T_1) states of an excited photoinitiator molecule:

$$I \xrightarrow{\ h\nu\ } I^*(S_1) \qquad \xrightarrow{\ ISC\ } \qquad I^*(T_1) \tag{4.132}$$
$$\downarrow \qquad\qquad\qquad\qquad \downarrow$$
$$R_1 \cdot + R_2 \cdot \qquad\qquad R_1 \cdot + R_2 \cdot \tag{4.133, 4.134}$$

where ISC is the intersystem crossing (cf. Sect. 2.3).

The photocleavage efficiency of initiator depends on:

(i)   an efficient population of the reactive excited singlet (S_1) and or triplet (T_1) states, which requires desirable absorptivity of photoinitiator, and a high efficiency of the intersystem crossing (ISC) process;

(ii)   deactivation of excited singlet (S_1) and or triplet (T_1) states by radiative processes (fluorescence and/or phosphorescence) and/or radiationless processes;

(iii) deactivation of triplet (T_1) state by oxygen.

Mechanisms of photocleavage of different photoinitiators have been discussed elsewhere in a number of publications [241–243, 352, 409, 455, 596–600, 606].

In Table 4.3 examples of photoinitiator-polymer systems are shown.

The overall efficiency of photoinitiation of polymer degradation depends upon:

(i) the fraction of incident UV (light) radiation absorbed by the photoinitiator which depends on the absorption spectrum of the photoinitiator and wavelength of radiation used;
(ii) photocleavage efficiency of the photoinitiator;
(iii) reactivity of radicals formed with a polymer (i.e. efficiency of hydrogen abstraction);
(iv) the initiator molecule, or any of its photocleavage products (radicals formed) should not function as a chain transfer or termination agent in photodegradation.

There is an optimal concentration of photoinitiator which can cause chain scission and/or crosslinking reactions. An increase in the photoinitiator concentration causes the following effects:

(i) more effective and rapid degradation of a polymer due to more radicals produced from the photocleavage of a photoinitiator;
(ii) once the threshold photoinitiator concentration is exceeded, the rate of degradation is essentially independent of photoinitiator concentration — this threshold photoinitiator concentration is dependent on the extinction coefficient (ε) and the samples thickness (l);
(iii) further increase of photoinitiator concentration may cause free radicals from photocleavage to dominate the termination reactions;
(iv) compatibility of the photoinitiator with the polymer matrix;
(v) presence of additives (antioxidants, thermostabilizers, photostabilizers, inihibitors, etc) which can scavenge radicals formed from photocleavage.

In the photoinitiated degradation it is important to find the most efficient photoinitiator and determine its optimum concentration for the most effective initiation of polymer degradation.

4.31 Photoinitiated Degradation of Polymers by Dyes

Most practically utilized polymers and plastics are dyed. Dye photoinitiated degradation of polymers is a phenomenon commonly known in industry as phototendering, and has been reviewed in a number of publications [14, 31, 492, 632].

A dye molecule (D) absorbs light and is excited to the excited singlet (^1D) and/or triplet (^3D) states (denoted here as D^*):

$$D \longrightarrow D^* . \tag{4.135}$$

Table 4.3. Examples of various types of photoinitiators for accelerated degradation (chain scission and/or crosslinking) of different polymers

Photoinitiator	Fragmentation reaction	Polymer	References
Acetophenone derivatives		Polyethylene EPDM B	557 90, 96
Benzoin derivatives		Polyethylene Poly(methyl methacrylate) Poly (vinyl chloride) Polystyrene	557 493 358 701
Chloronitroso-compounds		Polyisoprene	592, 593, 595 614

Photoinitiator	Hydrogen abstraction reaction	Polymer	References
Benzophenone		Polyethylene	136, 417, 419, 557, 801, 823
		Polypropylene	52, 334, 335, 416, 418, 419
		Poly(ethylene-co-propylene)	91, 94, 288
		Poly(vinyl acetophenone)	794
		Poly(ethylene glycols)	54, 163, 803
		Poly(vinyl chloride)	342, 358, 558–560
		Polystyrene	287, 450, 503, 701, 760, 814
		Poly(α – methylstyrene)	352, 814
		Polybutadiene	299
		EPDM B	91, 94–97
		Polynorbornene	808
		Cellulose	499, 500
Quinones		Polyethylene	222, 348, 589, 749, 750, 824, 825
		Polypropylene	34
		Poly(vinyl chloride)	703
		Polystyrene	617
		Polyamides	32, 215
		Cellulose	215

The photoactivated dye molecule (D^*) may abstract a hydrogen atom from a polymer (PH) molecule and produce a polymer alkyl radical ($P\cdot$) and protonated dye radical ($DH\cdot$):

$$PH + D^* \longrightarrow P\cdot + DH\cdot \, . \tag{4.136}$$

In the presence of water (moisture) (in particular vat dyes) an additional reaction may occur:

$$D^* + H_2O \longrightarrow DH\cdot + HO\cdot \tag{4.137}$$

$$DH\cdot + O_2 \longrightarrow D + HO_2\cdot \, . \tag{4.138}$$

Hydroxy ($HO\cdot$) and/or hydroperoxy ($HO_2\cdot$) radical may further abstract a hydrogen atom from a polymer (PH) molecule (cf. Sect. 4.16).

The deactivation of the photoactivated dye molecule (D^*) may produce a semioxidized radical cation ($D\cdot^+$) from the dye [435, 572, 573]:

$$D^* \longrightarrow D\cdot^+ + e^- \tag{4.139}$$

or a semireduced radical anion ($D\cdot^-$) from the dye:

$$D^* + e^- \longrightarrow D\cdot^- \tag{4.140}$$

$$D^* + HO^- \longrightarrow D\cdot^- + HO\cdot \, . \tag{4.141}$$

The disproportionation reaction may cause a recovery of dye:

$$D\cdot^+ + D\cdot^- \longrightarrow 2D \, . \tag{4.142}$$

Both dye radicals may also participate in the following reactions:

$$D\cdot^+ + DH\cdot \longrightarrow 2D + H^+ \tag{4.143}$$

$$D\cdot^+ + OH^- \longrightarrow D + HO\cdot \, . \tag{4.144}$$

The dye radical anion ($D\cdot^-$) often has a strong tendency to take a proton and/or exist in equilibrium with it:

$$D\cdot + H^+ \rightleftharpoons DH\cdot \, . \tag{4.145}$$

Both $D\cdot^-$ and $DH\cdot$ represent semireduced forms of the dye molecule in the ground state (S_0), and both are included in the term semiquinone. Dye radicals ($DH\cdot$) are very reactive and produced a colourless product (DH_2) known as the leucoform of the dye by a disproportionation reaction:

$$DH\cdot + DH\cdot \longrightarrow D + DH_2 \tag{4.146}$$

or hydrogen atom abstraction:

$$DH\cdot + PH \longrightarrow DH_2 + P\cdot \tag{4.147}$$

and in oxygenated solutions produce the oxygen radical ion ($O_2\cdot^-$):

$$DH\cdot + O_2 \longrightarrow D + H^+ + O_2\cdot^- \, . \tag{4.148}$$

In the presence of water, an electron (e^-) can be solvated (e^-_{aq}) and can cause the following reactions:

$$e^- \xrightarrow{\text{H}_2\text{O}} e^-_{aq} \tag{4.149}$$

$$e^-_{aq} + O_2 \longrightarrow O_2 \cdot^- \tag{4.150}$$

$$O_2 \cdot^- + H_2O \longrightarrow HO_2 \cdot + HO^- \tag{4.151}$$

$$2\,HO_2 \cdot \longrightarrow H_2O_2 + O_2 \tag{4.152}$$

$$HO_2 \cdot + e^-_{aq} \longrightarrow HO_2 \cdot^- \tag{4.153}$$

$$HO_2 \cdot + H_2O \longrightarrow H_2O_2 + OH^- \tag{4.154}$$

$$2\,O_2 \cdot^- + 2H^+ \longrightarrow H_2O_2 + O_2 \ . \tag{4.155}$$

The excited dye molecule can also produce singlet oxygen ($^1O_2$) by the energy transfer mechanism (cf. Sect. 4.29):

$$D^* \ (T_1) + O_2 \longrightarrow D \ (S_0) + {}^1O_2 \ ({}^1\Delta_g) \ . \tag{4.156}$$

In Table 4.4 several common dyes which efficiently produce $^1O_2$ are shown. Their triplet energies (E_T) range from 30–56 kcal mol^{-1}.

All of these reactions may cause dye fading. The mechanisms of dye fading are complex and often dependent upon the structure of the dye and the chemical and/or physical nature of the polymer [12, 14, 435, 826, 831].

"Phototendering" occurs when a dye accelerates the breakdown in the molecular structure of a polymer. In the case of tendering dyes in viscose and nylon-6 the following conclusions have been drawn [14, 31, 37]:

(i)   the active tendering dyes are mainly among the yellow and orange dyes, although there are some reds and blues that are active and there are also some non-tendering yellow and oranges ones;
(ii) the presence of oxygen is essential for significant tendering to occur;
(iii) the presence of water vapour has a marked accelerating effect on the tendering rate, particularly with cellulose polymers;
(iv) the nature of the polymer is important – fibres made from viscose rayon, cotton and nylon readily degrade, while with wool and cellulose esters degradation is virtually non-existent;
(v)   the degree of aggregation of the dye is important but the overall effect differs with different types of dye and the after-treatment of the polymer;
(vi) finally, and this factor is probably one of the most important to the dye manufacturer, there is obvious correlation between chemical structure and tendering activity.

The effect of phototendering can be decreased through treatment of a polymer with stabilizers.

Table 4.4. Triplet state energies of different dyes [582]

Dye	Triplet energy E_T (kcal mol^{-1})
Fluorescein	47.2
2,7-Dichlorofluorescein	46.0
Eosin B	45.5
Eosin Y	45.5
Rose Bengal	42.0

Table 4.4. (Continued)

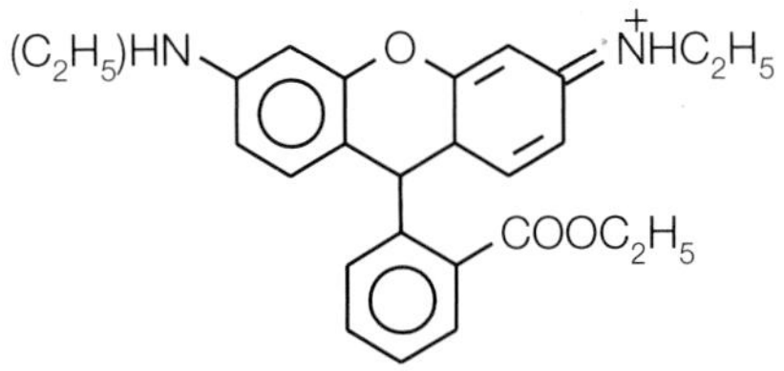

Rhodamine 6G   42.0

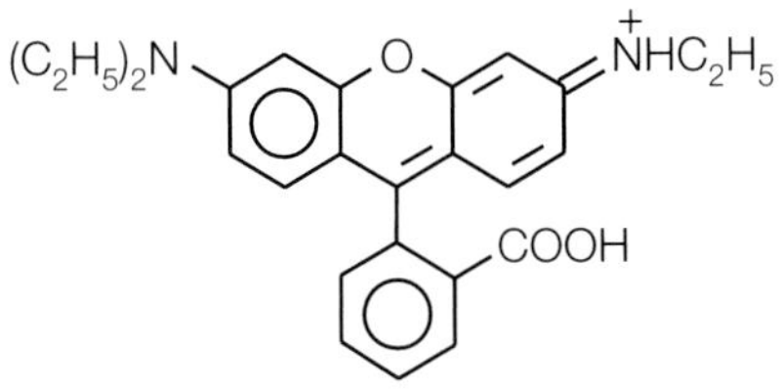

Rhodamine B   43.0

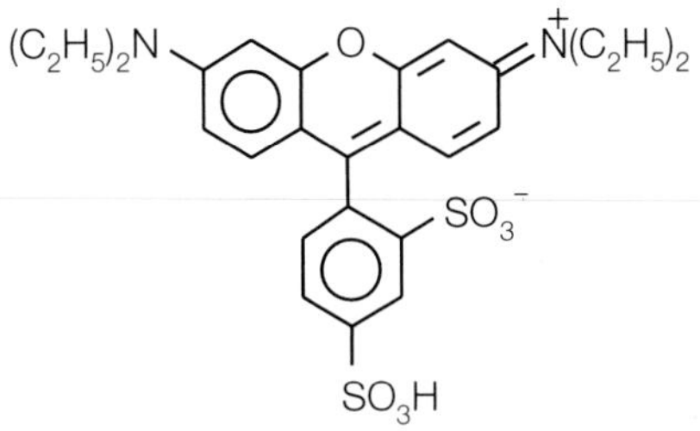

Sulphorhodamine B   41.8

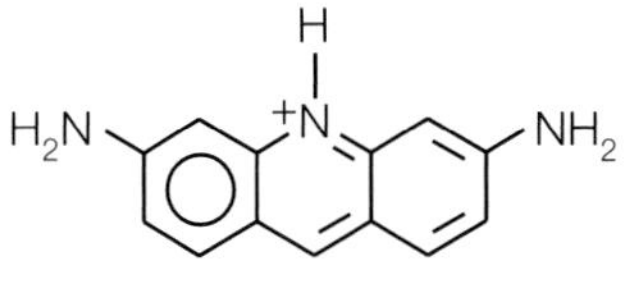

Proflavine   51.1

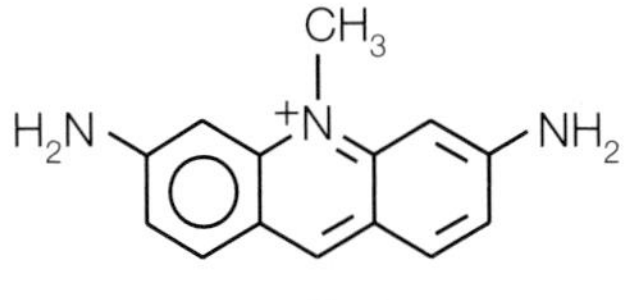

Acriflavine   51.1

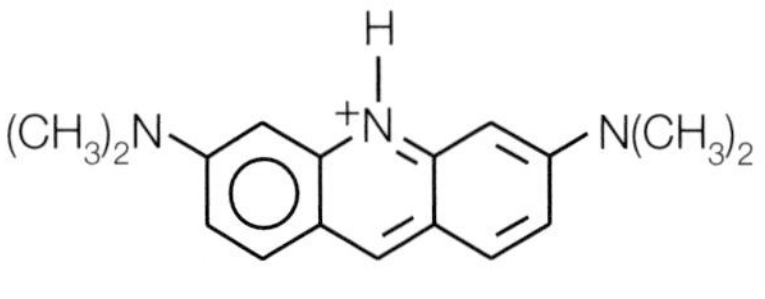

Acridine Orange

Table 4.4. (Continuedd.)

(CH$_3$)$_2$N — [structure] — $\overset{+}{N}$(CH$_3$)$_2$ Methylene Blue	32.0
H$_2$N — [structure] — $\overset{+}{N}$H$_2$ CH$_3$ Crystal Violet	42.0
CH$_3$ [structure] CH$_3$ Lumichrome	55.4
CH$_3$ [structure] CH$_3$ CH$_3$ Lumiflavine	50.5
CH$_3$ [structure] CH$_3$ CH$_2$ HO—CH HO—CH HO—CH CH$_2$OH Riboflavine	50.0

4.32   Photoinduced Electron Transfer

The basics of electron-transfer theory (introduced and developed by Marcus [476–482]) have been reviewed elsewhere [88, 114, 481, 482, 510].

Photoinduced electron transfer may occur between excited (S_1 or T_1) states of donor (D^*) or acceptor (A^*) and their ground states (D or A):

$$D^* + A \longrightarrow D^+ + A^- \tag{4.157}$$

$$D + A^* \longrightarrow D^+ + A^- \ . \tag{4.158}$$

These reactions occur by three steps.

(i)   The excited D^* (or A^*) and A (or D) diffuse together to form an outer-sphere precursor excited complex $(D–A)^*$:

$$D^* + A \longrightarrow (D-A)^* \tag{4.159}$$

$$D + A^* \longrightarrow (D-A)^* \ . \tag{4.160}$$

The precursor complex can also be formed in ground states of donor and acceptor and next by UV/visible radiation excited to (S_1 or T_1) states:

$$D + A \longrightarrow (D-A) \xrightarrow{\ h\nu\ } (D-A)^* \ . \tag{4.161}$$

(ii) In the second step, the excited precursor complex undergoes re-organization towards a transition state in which electron transfer takes place to a successor complex (D^+A^-):

$$(D-A)^* \rightleftharpoons D^+A^- \ . \tag{4.162}$$

(iii) In the third step, the successor complex dissociates to form the product ions D^+ and A^-:

$$D^+A^- \longrightarrow D^+ + A^- \ . \tag{4.163}$$

There are only a few examples of photodegradation of polymers by photo-induced electron transfer mechanism. For example, photodegradation of poly(α-methylstyrene) [389] and poly(2-phenyl-butadiene) [388], and photodegradation of partially fluorinated polyimides [153].

Photoinduced electron transfer plays a more important role in the photo-initiated hybrid photoinitiators systems (e.g. quinonesamines) as in the dye photoinitiated degradation of polymers. The photodegradation of natural polymers may also occur by the electron transfer mechanism.

5 Physical Factors Which Influence Photodegradation

5.1 Effect of Free Volume

Amorphous polymeric matrices provide a control of the amount of free volume available for chemical reactions [819].

The "free volume" (V_f) of a polymer matrix is defined by

$$V_f = V_T - V_0 \tag{5.1}$$

where V_T is the total volume of the matrix at temperature T (K), and V_0 is the theoretical molecular volume of the most dense packing of the matrix molecules at 0 K.

The free volume can be represented as the difference in two curves – the specific volume curve (Fig. 5.1, solid line) and the curve representing the expansion of the polymer chain with temperature based on bond length (Fig. 5.1, dotted line).

At absolute zero (0 K) there should be, in principle, no free volume, but in fact the occupied volume (V_0) includes not only the van der Waals radii but also a contribution from defects in the perfect crystalline lattice. As the temperature is raised, the amount of free volume increases, this being associated with the thermal vibrational motion. At sufficient free volume, side groups in a polymer chain (e.g. phenyl rings in polystyrene) start to move at a given frequency (the temperature designed by T_γ in Fig. 5.1). As the temperature is raised still further, additional free volume is available and at the point indicated by T_g ("glass transition temperature") a change in the slope of the specific volume is observed (Fig. 5.1) which is due to the occurrence of motion of very long segments of the polymer chain consisting of units of from 20 to 40 carbon atoms.

A polymer matrix provides a relatively continuous increase in the volume available for reactions to take place as the temperature is increased. Some of the motions which are dependent on the free volume are:

(i)   rotations of side groups, e.g. methyl, phenyl and others;
(ii)   formation of cyclohexane ring configuration (e.g. Norrish type II process, cf. Sect. 4.23);
(iii) diffusion of a small molecule (e.g. oxygen, or free radical like hydrogen or methyl radical) to a macromolecule where reaction can take place [151, 152, 471, 829];
(iv)   diffusion of large molecules containing donor and acceptor groups (this motion requires the largest amount of free volume to occur).

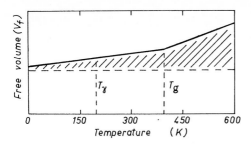

Fig. 5.1. Specific volume and "free volume" (V_f) (*dashed*) of a given polymer

The specific free volume (V_f) and the glass transition temperature (T_g) (for many polymers, e.g. polystyrene) depend linearly on the inverse of the number-average molecular weight ($\overline{M}_n$) [445, 708]:

$$V_f = V_0 + \frac{a}{\overline{M}_n} \tag{5.2}$$

and

$$T_g = T_g(\infty) - \frac{b}{\overline{M}_n} \tag{5.3}$$

where a and b are constants which are independent of the molecular weight.

The chain ends contribute to more free volume than the monomeric units present in the centre of the chain, and the molecular-weight dependence of the free volume is associated solely with the chain-end contributions [248].

The number of chain ends cm^{-3} is given by

$$N_c = \frac{2\rho N_A}{\overline{M}_n} \tag{5.4}$$

where ρ is the density of the sample, N_A is Avogadro's number (i.e. the number of molecules in a mole, $N_A = 6.02217 \times 10^{23}$ molecules mol^{-1}).

If one chain contributes $V(cm^3)$ to the free volume, then the free volume associated with the chain end (in cm^3) is

$$V_{fc} = \frac{2V\rho N_A}{\overline{M}_n} \quad . \tag{5.5}$$

The free volume associated with the chain ends corresponds to the increase of free volume when the molecular weight is varied from $\overline{M}_n$ to an infinite value. The expansion coefficient of the free volume (α_f) is

$$\alpha_f = \alpha_r - \alpha_g \tag{5.6}$$

where α_r and α_g are the expansion coefficients of the rubbery and glassy polymers, respectively.

The equation relating these quantities are

$$V_{fc} = \alpha_f[T_g(\infty) - T_g(\overline{M}_n)] = \frac{2V\rho N_A}{\overline{M}_n} \tag{5.7}$$

and therefore

$$T_g(\overline{M}_n) = T_g(\infty) - \frac{2V\rho N_A}{\overline{M}_n f} \quad . \tag{5.8}$$

The V value can be measured by dilatometric methods [782].

Breaking of polymer bonds under irradiation causes fragments to occupy more volume than the reactant, thereby causing formation of strains and stresses which can be responsible for the formation of micro-cracks and damage of irradiated material. In glassy polymers the free volume accounts for approximately 10% of the total volume [708]. If the chain scission is extensive, just a 10% increase in product volume is almost enough to completely fill up the free volume present in the polymer sample. If the initial molecular weight of photoirradiated polymer is high then the available free volume is relatively small. The chain-end-associated free volume only accounts for, at most, 3% of the total polymer volume (for a total free volume of 10%). As the chain-scission reaction proceeds, the molecular weight of the polymer decreases and therefore the chain-end-associated free volume increases, which effectively plasticizes the polymer glass. Chain-scission of a low molecular weight polymer will plasticize more than a high molecular weight material.

These volume recoveries occur by segmental motion [157]; as such, they are independent of sample size [102] and have a time-relaxation spectrum that depends strongly on the molecular weight of the sample ($\tau \propto M^3$). Each polymer is characterized by a molecular weight (M_e) below which this relaxation time changes dramatically (the entanglement effects are small) [708]. When the free volume reaches a critical value, the stresses developed in a polymer matrix cause the mechanical damage to an irradiated sample.

5.2   Effect of the Glass Transition Temperature

Three thermal transition temperatures in a large sense determine the physical character of a polymer (Fig. 5.2):

(i)   the glass transition temperature (T_g);
(ii) the crystalline melting temperature (T_m);
(iii) the polymer decomposition temperature (T_c).

Polymeric rotational freedoms and intermolecular bonding largely determine T_g and T_m temperatures.

Below the glass transition temperature (T_g), chain segments are frozen in fixed position in a disordered quasilattice. Some molecular movements of chain segments take place in the form of vibrations about a fixed position. A diffusional rearrangement of the segmental position is less probable. With increasing temperature, the amplitude of segmental vibrations increases. In the transition state, chain segments have sufficient energy to overcome the secondary intramolecular bonding forces. Chain segments or chain loops may perform rotational and transitional motions. In the rubbery state the seg-

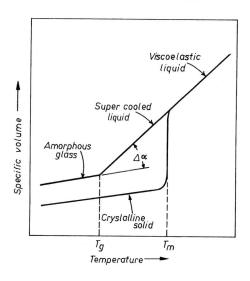

Fig. 5.2. Effect of the glass transition (T_g) and crystalline melt transition (T_m) upon the specific volume of polymers

mental motions are very rapid, whereas the molecular motion (the motion of the entire molecule) is restricted by chain entanglements. With increasing temperature the degree of entanglements decreases and the molecular slip increases, but the sample still shows some elasticity.

Above the glass transition temperature (T_g) the slip of entire molecules is predominant, and the elasticity of the sample disappears. The conformational freedom of the polymer chain can be comparable to that in solution (liquid state) and consequently has equivalent chemical reactivity.

Photoreactions which require very little change in the geometry of the excited state and reactants should proceed equally well in solid glassy polymer matrices and in solution. Dissociation of free radical pairs can be relatively efficient in the solid-state if one component is a small free radical but will be significantly inhibited if both components are polymeric radicals.

Bimolecular reactions which require the diffusion of a small molecule reagent to a species in a polymer matrix will depend on both the diffusion constant and the solubility of the material in the matrix. Diffusion in solid glass matrices, particularly of small molecules (or radicals), is much higher than would be predicted from the bulk viscosity of medium. Solid polymers generally have internal viscosities only two to three orders of magnitude less than those for simple liquids such as benzene or hexane, so that under suitable conditions quite efficient biomolecular reactions can be induced to occur by diffusional processes in polymeric materials [320].

In amorphous and partially crystalline polymers, the glass transition temperature (T_g) plays a role, since the mobility of the chains and radicals increases above T_g and, simultaneously, oxygen diffusion occurs more rapidly. The effect of this on photo-oxidative degradation rate depends on the material and the specific degradation process [761].

Reactions which can be considered to be associated with caged radicals, such as the photo-Fries rearrangement (cf. Sect. 4.28), will require very little free volume and can be expected to be quite efficient in solid polymers below the glass transition temperature (T_g), whereas photochemical processes such as the Norrish Type II process are substantially reduced in glassy polymers below the glass transition temperature (T_g), unless the geometry of the cyclic six-membered ring is particularly favoured by steric factors in the chain, so that the most stable conformation corresponds to that required for reaction [320].

5.3   Effect of Crystallinity

Many polymers are partially crystalline or crystallizable. The presence of such crystalline regions affects the chemical and physical properties of a polymer.

Crystallinity imposes certain constraints on the course of the photo (oxidative)-degradation of polymers. These constraints do not, in general, apply to the absorption of UV(light) irradiation and formation of excited singlet and/ or triplet states, but rather to subsequent secondary reactions, e.g. nature and stability of radicals, diffusion of oxygen to these radicals, etc.

The presence of crystalline regions restricts the mobility of the chain segments in the noncrystalline regions. In a sense, a crystallite behaves similarly to a chemical crosslink. The greater the density of crystallites, the less will be the mobility of chains in the amorphous regions. This will have the effect on the radical recombination process.

Such crystallinity means that radicals which would be formed within a crystalline region are virtually trapped, and may have lifetimes of many months, even years, under certain commonly attainable conditions.

A much more subtle effect which crystallinity may have on the overall interaction of UV(light) radiation with polymers is that it may provide a driving force for certain types of radical formation over others. The driving force derives from the forces which maintain the integrity of the crystal lattice. The abstraction of a hydrogen atom from the polymer molecule would be favoured over the rupture of a carbon-carbon backbone bond. This is so for two reasons.

(i)   First, the small hydrogen atom can readily diffuse through the crystal lattice and remove itself from the zone of interaction with its partner macroradical. This cannot be done by either of the two polymer radicals ($P \cdot$) formed from a backbone carbon-carbon bond rupture, with the result that this type of a chain breaking will lead to two very closely spaced free radicals which will probably interact to reform the original bond by a termination reaction.

(ii) Second, the radical resulting from the first type of bond rupture, upon rearrangement of the bond angles around the carbon atom associated with the free radical, can be more easily accommodated within the geometry of the crystal lattice than can the two adjacent but separated radicals resulting from the latter type of bond dissociation.

By this close interplay of crystal geometry requirements and crystal potential energy, crystallinity may influence the course of the reaction in a suitable fashion. For example, a glassy structure (an amorphous polymer below its glass transition temperature T_g) is also fertile ground in which to build up a trapped free-radical population.

Most photochemical reactions cannot occur in the crystalline state due to [317, 320]:

(i)   the possible delocalization of the excitation through the crystal;
(ii) the rigidity of the crystal lattice;
(iii) the high symmetry (a high degree of order), very close packaging of the polymer molecules, and lower (or even lack of) free volume (imposed by the rigidity of crystal lattice) with a consequent reduction of chemical reactivity;
(iv) the lack of oxygen in the crystalline lattice.

Unless the photon energy is sufficient to cause local increase of temperature and melting of the crystal lattice to provide the necessary free volume, it seems unlikely that photochemical processes, particularly of the type requiring large rearrangements of molecular structure, can occur efficiently in crystalline regions of a polymer.

Oxidation of semicrystalline polymers such as polyolefins (polyethylene and polypropylene) is generally considered to occur within the amorphous region which can be treated as a boundary phase of the neighbouring crystalline regions (Fig. 5.3) [275, 759]. The molecules which connect crystallites through amorphous regions are scissioned in the oxidation process, resulting in a decrease in elongation and changes in other physical properties. At later stages of oxidation, when many chains in the amorphous phase and at the crystalline boundary are destroyed, samples exhibit brittleness upon external stress.

The photostability of quenched linear low density polyethylene (LDPE) is superior to that of annealed linear low density polyethylene. Quenched

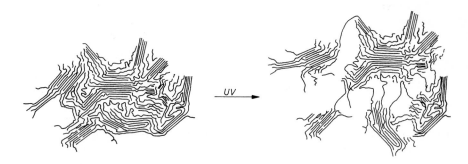

Fig. 5.3. Schematic presentation of oxidation of the amorphous region in a semicrystalline polymer causing chain scission

polyethylene has crystallites smaller than those of annealed polyethylene. However, the degree of crystallinity does not depend upon the method of film preparation. Consequently, quenched linear low density polyethylene contains a greater number of tie molecules and is able to attain a higher build-up of oxidation products without causing serious damage to the tie molecules [762].

5.4   Localization of the Oxidation Processes in Semicrystalline Polymers

It is generally accepted that photo-oxidation does not occur in the crystalline part of a polymer. It has been shown that crystalline regions of polyethylene are impenetrable to oxygen [501, 680]. Direct measurements of oxygen solubility in polypropylene with different degree of crystallinity also show that oxygen solubility in the crystalline fraction of polypropylene is at least one order of magnitude lower than in its amorphous fraction [402].

Impurities facilitating the oxidation initiation (e.g. traces of polymerization catalysts, carbonyl, hydroperoxide and unsaturated groups) are displaced into the amorphous zones during polymer crystallization. This will also lead to a higher initiation rate in the amorphous phase compared with the crystalline fraction.

Stabilizers added to a polymer are distributed in the amorphous phase only.

5.5   Chemicrystallization Process

During degradation of some polymers (e.g. polyesters) density of samples increases due to a chemicrystallization process.

A chemicrystallization process can be the result of two processes:

(i)   chain-ends occurring at a higher rate than a random in-chain process, enhancing the degree of water-extractable products [61];
(ii) chain-scission of previously entangled chains in the amorphous regions giving them sufficient mobility to crystallize [563, 804].

Crystallinity changes during the degradation of polymers are particularly important in relation to other chemical changes like oxidation, chain scission, crosslinking and hydrolysis.

The initial crystallinity of biaxially oriented polyester bottle materials has an important influence on the rate of hydrolytic degradation of the polymer because the crystallites act as barriers to moisture and oxygen diffusion [16].

Crystallization and main chain scission are not necessarily related in degradation and hydrolysis reactions and therefore interpretations of stability by mechanical testing such as tensile properties may not correlate.

5.6   Effect of Polymer Molecular Weight

The molecular weight of the polymer can affect the photoreactions occurring in the solid state polymer matrix. Photoreactions occurring in organic solids are diffusion controlled. A chemical bond in a macromolecule breaks irreversibly if the two radicals diffuse apart from each other by a distance (the diameter of the recombination sphere) during their effective lifetimes. Small chemical radicals move by a simple Fick's law-type diffusion, but polymeric radicals move by a process that was called "reptation" [194]. This motion is much slower, so that a bond present in a pendant group is easier to break than the same bond which is present in a polymer backbone. Theoretical studies of intramolecular [556] and intermolecular [195] diffusion processes in condensed polymer melts showed that for a short reaction time (t $< \mu s - $ ms), the molecular displacements are smaller than the distance between two entanglement points, i.e. the reaction is not affected by the reptational diffusion. Therefore fast diffusion-controlled photoreactions in "high-value" polymer (N $<$ N$_e$, where N$_e$ is the number of monomers per entanglement point) should be nearly molecular-weight independent. However, these processes are still diffusion controlled and very much less efficient than in low molecular weight compounds.

Longer chains are more subject to oxidative attack and chain rupture than shorter chains. The statistically more probable scissioning of longer chains has a greater impact on changes in weight-average molecular weight ($\overline{M}_w$) than on number-average molecular weight ($\overline{M}_n$). Chain scissioning affects not only molecular weight but also mechanical properties. The rate of formation of oxidation products is faster in the crosslinked polyethylene [756, 762].

The stoichiometry of bond breaking process indicates that only one bond break cuts the molecular weight in half, regardless of the initial molecular weight of the polymer. Polymers with very large molecular weights will therefore require fewer breaks per unit weight than polymers with low molecular weight in order to reduce their molecular weights by the same amount. In many papers it is common practice to plot either the molecular weight (or the intrinsic viscosity [η] or more properly [η]$^{1/\alpha}$) as a function of the reaction time (Fig. 5.4). However, it is not completely correct if one is considering the stoichiometry of bond breaking. For that reason, the only appropriate method of plotting such data is to plot $1/\overline{M}_n$ or the reciprocal of [η]$^{1/\alpha}$ as a function of either time or the radiation absorbed, as shown in Fig. 5.5 [317].

If the process involves a constant rate of bond breaking, then a straight line will be observed, the slope of which can be related either to the rate constant or the quantum yield of thermal- or photochemical processes, respectively. Either upward or downward curvature of plots of the form of Fig. 5.5 can then be accurately interpreted in terms of changes in either the mechanism or rate of some of the fundamental processes involved in degradation. Because of the difficulty in interpreting data plotted in the form of Fig. 5.4, the use of such plots should be avoided [317].

The plot of 1/M will not be linear if $\overline{M}_n$ is the number-average molecular weight (M). If any other average molecular weight is used, then it is necessary

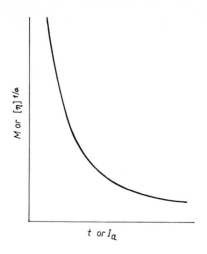
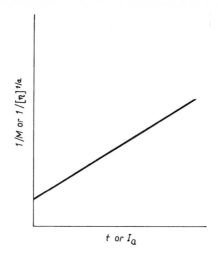

Fig. 5.4. Plot of changing of molecular weight (M) or viscosity ($[\eta]$) vs time of irradiation (t) or radiation absorbed (I_a)

Fig. 5.5. Plot of changing of inverse molecular weight (M) or viscosity ($[\eta]$) vs time of irradiation (t) or radiation absorbed (I_a)

to determine the changes in distribution which occur in the course of the degradation process and convert the average to $\overline{M}_n$.

5.7 Effect of the Formation of Hydrogen Bonds

"Hydrogen-bonding" is a particular electrostatic type of interaction. Hydrogen is able to form a link between two atoms, most strongly with fluorine, oxygen and nitrogen, which enhances the total energy.

The increase in the amount of photo-oxidation products can be accompanied with the reduction of the segmental free volume since the oxidation products being formed (hydroperoxy, hydroxy, carbonyl or carboxyl groups) interact in an unusual way through attractive dipole–dipole interactions and especially through specific hydrogen bonding between acidic hydrogens and basic oxygen groups. Such inter- and/or intra-molecular interactions lead to a more rigid polymer chain and thus to a decrease in the size of the polymer segment able to move at a particular temperature. However, crosslinking and other reactions can also lead to a reduction of the chain mobility.

During the formation of hydrogen bonds, the inter-chain distances are reduced ("pulling effect") (Fig. 5.6). For example, photo-oxidation of polyimide film causes its gas (oxygen/nitrogen) permeability to decrease and selectivity to increase due to a pulling effect [495].

Although the energies of hydrogen bonds are weak (4–12 kcal mol^{-1}) in comparison to covalent bonds (of the order of 90 kcal mol^{-1}), this type of interaction is large enough to produce appreciable frequency and intensity in the vibrational IR spectra. In fact, the disturbances are so significant that mid-

IR and Raman spectroscopy have become one of the most informative sources of criteria for the presence and strength of hydrogen bonds [407, 505, 698–700, 783].

5.8   Cage Recombination

The "cage effect" is related to encirclement of the reaction products by the solvent and is typical in solution photochemistry. However, it can also be applied in an amorphous polymer matrix, where the reaction products are encircled by macromolecules.

High liquid density restricts the molecular movement. As a consequence of high density in the liquid phase, molecules must spend a relatively large time oscillating around a position in the midst of solvent molecules, i.e., colliding with their nearest neighbours, and only occasionally experiencing a net displacement. Photo-(thermal-) dissociation of the excited molecule (chromophoric group) produces a pair of radicals (often called "primary radicals"). Collisions of these radicals within the solvent cage will lead to "primary recombination", a process that competes with diffusion out of the cage. In this process the free radicals may never attain a separation of as much as a molecular diameter and recombination takes place in a period that is longer than a vibration (10^{-13} s) and less than the time between diffusive displacements (10^{-11} s). When free radicals undergo diffusive separation, i.e. escape from the cage, there is still the probability that they will reencounter each other later, after a number of random displacements ("diffusive recombination"). Radicals failing to undergo primary and diffusive recombination must diffuse into the bulk of the solution and enter into steady state reactions, e.g.

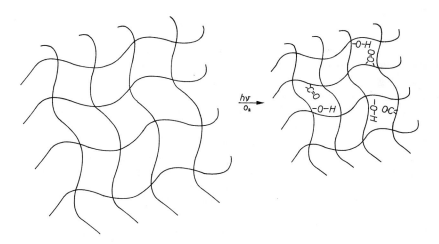

Fig. 5.6. "Pulling" the polymer chains closer together through hydrogen bonding of polar groups [495]

(i)   abstraction of hydrogen from the same and/or neighbouring macro-molecule with formation of polymer alkyl (P·) radical;
(ii) participation in the termination reaction with any available polymeric radical (P·, PO·, POO·).

The events that take place during primary and diffusive recombinations cannot be described by conventional kinetics since the probability of pair recombination at this time is proportional to $t^{-3/2}$.

The dependence of the cage effect can be rationalized by means of the rates of free radicals formation (dissociation) (k_d) and primary recombination of free radicals formed (k_c):

$$\text{species absorbing radiation} \xrightarrow[k_d]{h\nu} R_1\cdot + R_2\cdot \quad \text{(dissociation)} \qquad (5.9)$$

$$R_1\cdot + R_2\cdot \xrightarrow[k_c]{} \text{product} \qquad \text{(recombination) .} \qquad (5.10)$$

The reciprocal of the "cage effect" (1/f)

$$\frac{1}{f} = \frac{k_c + k_d}{k_c} \qquad (5.11)$$

depends on the square root of the viscosity of polymer solution $(1/\eta)^{1/2}$.

The cage effect has been a subject of many theoretical predictions and models [103, 408, 457, 533, 535].

6 Kinetic Treatments of Degradation

6.1 Photodegradation in Isothermal and Non-Isothermal Conditions

Most photodegradation experiments are usually carried out in isothermal conditions, whereas polymers and plastics work in the environmental conditions within non-isothermal parameters. The rates of many degradation processes in polymers generally increase as the temperature rises.

The extent of the degradation process for non-isothermal measurements can be expressed by

$$-\frac{d\alpha}{dt} = k(1 - \alpha)^n \tag{6.1}$$

where α is the extent of any reaction monitored during the degradation, k is the rate constant and is thermally activated, following an "Arrhenius expression"

$$k = A\ e^{-(E_a/R_g T)} \tag{6.2}$$

where A is the pre-exponential factor, E_a is the activation energy, R_g is the gas constant, and T is the absolute temperature, and n is the order of reaction, and can be determined by a series of isothermal measurements at different temperatures.

The extent of the degradation process for the non-isothermal conditions (for the rising temperature) can be given by

$$\frac{d\alpha}{dT} = \frac{(1 - \alpha)^n}{b} A\ e^{-(E_a/R_g T)} \tag{6.3}$$

where $b = dT/dt$ is the heating rate.

The temperature-lifetime relationship is given by:

$$t = C\ e^{-(E_a/R_g T)} \tag{6.4}$$

where C is the temperature-independent constant.

For a particular activation energy, Eq. (6.4) reduces to

$$\log\left(\frac{t_1}{t_2}\right) = \frac{T_2}{T_1}\ . \tag{6.5}$$

An approximate rule drawn from reaction kinetics states that a temperature increase of 10°C will approximately double the reaction rate. This indicates the great significance of the temperature parameter for the ageing of polymers and plastics in the various climatic zones of the World.

6.2   Decay of Free Radicals in a Polymer Matrix

The photo-oxidative degradation involves the same radical intermediates – $P\cdot, PO\cdot$ and $PO_2\cdot$. In the solid state, restrictions in radical mobility lead to much shorter chain lengths than those found in the liquid phase [317]. The lifetimes of radical intermediates is in the order $P\cdot < PO\cdot < PO_2\cdot$. Free isolated $PO_2\cdot$ radicals in atactic polypropylene have a half-life of ~ 500 s [488]. However, many radicals will be generated as caged pairs and have very much shorter lifetimes because of rapid recombination (cf. Sect. 5.8).

The kinetics of decay of free radicals generated in a polymer matrix is usually interpreted in terms of the second-order equation

$$\frac{1}{[R]} - \frac{1}{[R_0]} = kt^{1/2} \tag{6.6}$$

where [R] and $[R_0]$ are the concentrations of the decaying radicals at time t and zero respectively, t is the time in which decay occurs, and k is the second-order rate constant.

The second-order rate of a diffusion controlled reaction between two radicals, based on the so-called Smoluchowski [705] boundary condition is given by [786]

$$\frac{1}{[R]} - \frac{1}{[R_0]} = k_D \left[1 + \frac{2r_0}{(\pi Dt)^{1/2}} \right] t \tag{6.7}$$

where k_D is a constant equal to $4\pi r_0 D$, D is the sum of the diffusion coefficients of the individual reacting radicals, and r_0 is the free-radical separation distance within which they react and outside of which the potential of unreacted radical is independent of position (Smoluchowski boundary condition). The radius r_0 may be considered to be the radius of the reaction cage.

Equation (6.7) can be converted to a linear form [203]

$$\left(\frac{1}{[R]} - \frac{1}{[R_0]} \right) \frac{1}{t^{1/2}} = k_D t^{1/2} + \frac{2r_0 k_D}{(\pi D)^{1/2}} \quad . \tag{6.8}$$

By plotting the left-hand side of Eq. (6.8) as a function of $t^{1/2}$ it can be readily seen whether a linear relation is obtained and, more importantly, whether $(1/[R] - 1/[R_0])/t^{1/2}$ extrapolates to a finite intercept at time t equal to zero, in which case the reaction is diffusion controlled (Fig. 6.1). An ordinary second-order reaction would give a zero intercept [203].

The term $2r_0/(\pi Dt)^{1/2}$ inside the bracket in Eq. (6.7) is usually negligible with respect to unity because in most liquid systems D is of the order

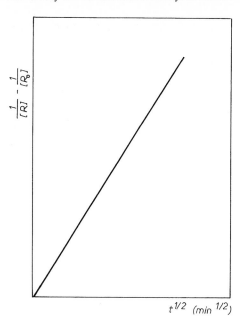

Fig. 6.1. Kinetic of decay of free radicals generated in a polymer matrix interpreted in terms of the second-order equation (Eq. 6.8)

$t^{1/2}$ $(min^{1/2})$

10^{-6} cm^2 s^{-1} and $r_0 \sim 10^{-7}$ cm, but if the reaction occurs in times of the order of micro- or nano-seconds, then $2r_0/(\pi Dt)^{1/2}$ will no longer be negligible and the reaction becomes diffusion controlled. However, if the diffusion constant is orders of magnitude smaller than 10^{-6} cm^{2}s^{-1} in reactions occurring in the solid state, there is the possibility that the term $2r_0/(\pi Dt)^{1/2}$ in Eq. (6.7) may be much greater than unity at all times up to 100 min. For unity being only 10% of $2r_0/(\pi Dt)^{1/2}$ at t = 100 min or 6×10^3 s, it is necessary that r_0 be as large as 4×10^{-7} cm and D as small as 3.4×10^{-19} cm^{2}s^{-1}. If unity can be neglected in Eq. (6.7), then the second-order rate constant (k) (Eq. 6.6) is

$$k = [2r_0 k_D/(\pi D)^{1/2}] = 8r_0^2(\pi D)^{1/2} . \tag{6.9}$$

Such reactions are associated with "time independent diffusion controlled reaction rate constants" [203].

Deviations from the second-order decay kinetic can be the result of:

(i)   non-diffusion-controlled process of free-radical decay;
(ii)   mechanism of free radical recombination.

Free radicals statistically dispersed in a solid polymer matrix are subjected to the motion of segments carrying the radical site [752–754]. Several types of radical chain motion like crank, crankshaft and kink participate in a free-radical decay [63]. The motion of polymer segments carrying the radical site depends on the free volume. In a new theoretical approach the decay of free radicals in solid polymers is based on a concept of voids existing in each polymer with the mobility induced by the motion of respective polymer segments [661].

6.3   Spin Trapping of Polymeric Radicals

"Spin trapping" is a reaction where short-lived low-molecular (R·) and/or polymeric radicals (P·) are scavenged by a diamagnetic low-molecular compound (S) to form a persistent radical adduct (R–S· or P–S·):

$$\text{R·(or P·)} + \text{S} \longrightarrow \text{R} - \text{S·(or P} - \text{S·)} \ . \tag{6.10}$$

In this reaction the diamagnetic scavenger (S) is called the "spin trap", and the resulting persistent radical (R–S· or P–S·) the "spin adduct". Unlike many short-lived radicals R· (or P·), the spin adduct R–S· is readily detectable by ESR even in liquid solution at higher temperature.

Although more than 100 different compounds have proven to be suitable spin traps, nitrones and C-nitroso compounds, which both give persistent aminoxyls upon radical addition:

$$\text{P·} + \text{R}_1 - \text{CH} \overset{+}{=} \text{N} - \text{R}_2 \longrightarrow \text{R}_1 - \text{CH} - \overset{\cdot}{\overset{|}{\text{N}}} - \text{R}_2 \tag{6.11}$$

$$\text{P·} + \text{R}_3 - \text{N} = \text{O} \longrightarrow \text{R}_3 - \overset{\cdot}{\overset{|}{\text{N}}} - \text{P} \tag{6.12}$$

are commonly used.

The structure of some typical nitrones used in spin trapping experiment are given below:

PBN   DMPO   GBBN

(MO) 3PBN   2- SSPBN   DBN

Next to nitrones, aliphatic and aromatic nitroso compounds are the most used spin traps [660]:

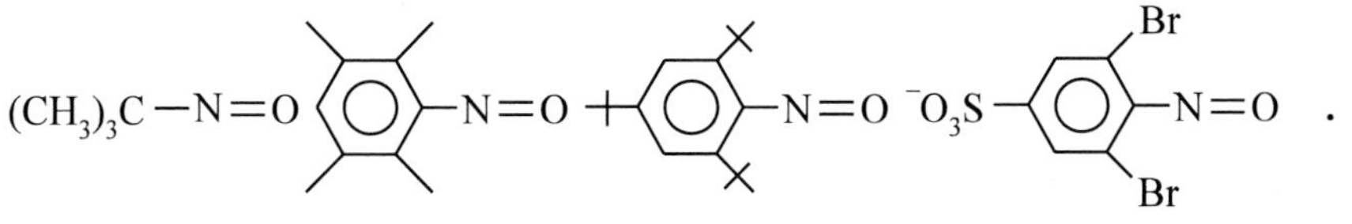

$$(CH_3)_3C-N=O \qquad \text{ND} \qquad \text{BNB} \qquad \text{SBNS}$$

MNP   ND   BNB   SBNS

The spin trap 2,3,5,6-tetramethylnitrosobenzene (ND) reacts with two different polyethylene alkyl radicals giving two types of spin-adduct (A) and (B) according to the following mechanisms [587, 588]:

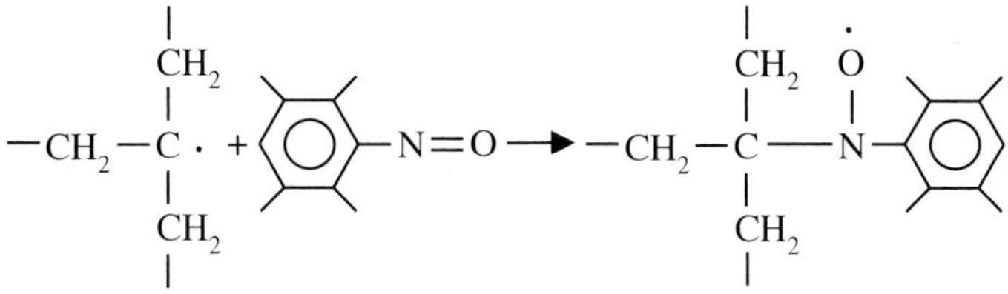

$$(6.13)$$

(A)

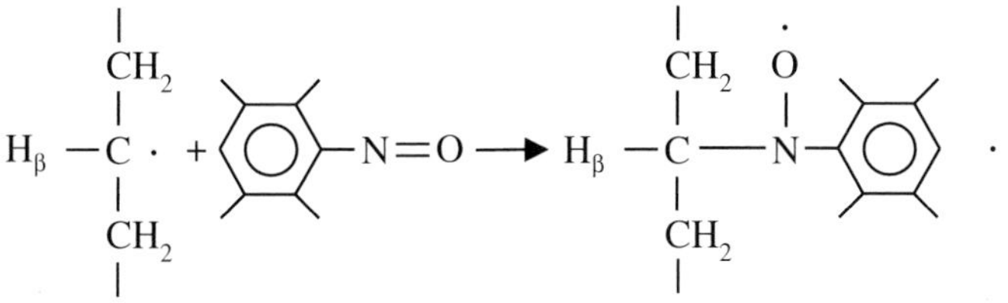

$$(6.14)$$

(B)

The ESR spectra of spin adducts (A) and (B) overlap each other (Fig. 6.2). The triplet is attributed to the tertiary carbon radical (A), whereas the triplet of doublets are assigned to a spin-adduct radical (B) where the original triplet is further split by the β-H of the secondary carbon radical adduct (B).

Another spin trap, 2,4,6-tri-*tert*-butylnitrosobenzene (BNB) has two trapping sites, one on the nitrogen to attach a non-bulky radical to form a nitroxide spin-adduct and another on the oxygen atom of the nitroso group to attach a bulky group to form an anilino spin-adduct [740, 741]:

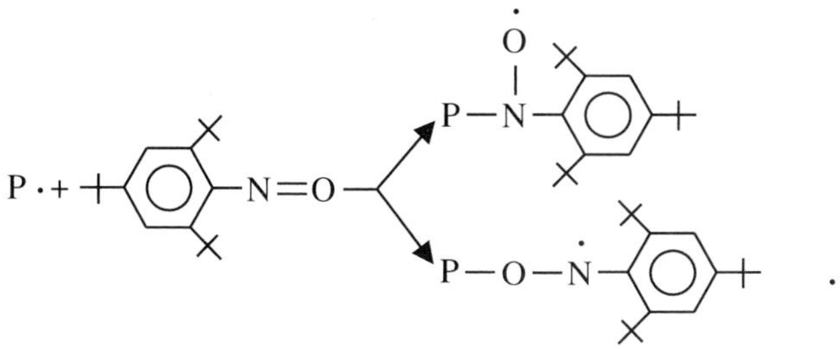

$$(6.15)$$

$$(6.16)$$

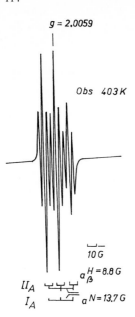

$g = 2.0059$

Obs 403 K

10 G

II_A

I_A

$a_\beta^H = 8.8 G$

$a^N = 13.7 G$

Fig. 6.2. ESR spectra of two overlapped spin adducts (A) I_A and (B) II_A formed between two different polyethylene alkyl radicals and the spin trap 2,3,5,6-tetramethylnitrosobenzene (ND) [587]

It is possible to distinguish between the anilino and the nitroxide radicals from the differences in the g values and the nitrogen splitting constants [587, 588].

Nitroxyl spin probes have also been used to obtain information on the photo-oxidation of isotactic polypropylene [425], Poly(ethylene oxide) and cellulose [338, 815].

Unlike nitrones, many nitroso compounds are dimers in the solid state. In order to act as spin traps, dissociation into monomers has to occur. This may lead to some difficulties in kinetic spin trapping experiments. However, because of steric hindrance by the bulky *tert*-butyl groups, 2,4,6-tri-*tert*-butylnitrosobenzene (BNB) exists exclusively as a monomer. The main disadvantage of using nitroso compounds as spin traps in photoreactions is their photochemical instability [137, 644].

An extensive compilation of spin adducts ESR data has been published recently [109]. Applications in photochemistry [332, 642, 643] and polymer research [113, 634] have also been reviewed.

6.4 Photodepolymerization

"Depolymerization" occurs as a chain reaction with successive splitting-off of monomer units from chain ends. This process is named the "unzipping" or "peeling-off" reaction.

The depolymerization mechanism depends on the initiation stage and can be divided into two types [130, 396, 463]:

(i)   the reaction starts by random breaks of internal bonds in chains;
(ii) the reaction begins at the polymer chain ends.

Depolymerization may further occur by a partial unzipping or by a complete unzipping.

"Polymer depolymerization" can be described in terms of the decreasing of the weight of polymer sample (w) during time (t):

$$w = N m \overline{DP} \tag{6.17}$$

$$\frac{dw}{dt} = m\left(N\frac{d\overline{DP}}{dt} + \overline{DP}\;\frac{dN}{dt}\right) \tag{6.18}$$

where N is the number of molecules, m is the molecular weight of the monomeric unit, and $\overline{DP}$ is the average degree of polymerization, related to the average molecular weight of the polymer ($\overline{M}$):

$$\overline{DP} = \frac{\overline{M}}{m} \;. \tag{6.19}$$

For random initiation followed by incomplete (partial) unzipping, and assuming that depolymerization is first-order with respect to sample weight (w), the following equations can be written:

$$\frac{dN}{dt} = k\frac{w}{m} \quad \text{and} \tag{6.20}$$

$$-\frac{dw}{dt} = kw\overline{Z} \tag{6.21}$$

where k is the rate constant of the depolymerization, and $\overline{Z}$ is the average unzipping length, i.e. numbers of mers formed from the unzipping of a chain.

Combination of Eqs. (6.20) and (6.21) followed by integration gives

$$\frac{\overline{DP}}{\overline{DP}_0} = \frac{\overline{Z}}{\overline{DP}_0}\left(\frac{1-C}{C\;\overline{DP}_0 + \overline{Z}}\right) = f\left(\frac{1-C}{C+f}\right) \tag{6.22}$$

where

$$f = \frac{\overline{Z}}{\overline{DP}_0} \tag{6.23}$$

and C is the fractional conversion, given by

$$C = \frac{w_0 - w}{w_0} \tag{6.24}$$

where index 0 is for undepolymerized samples.

For random initial depolymerization followed by complete unzipping, the average unzipping length ($\overline{Z}$) can be related to the average degree of polymerization ($\overline{DP}$) through a parameter (B), that is related to the polydispersity of the polymer sample:

$$\overline{Z} = \overline{DP}\;B \;. \tag{6.25}$$

Combining Eqs. (6.20), (6.21) and (6.25) into Eq. (6.18) with integration gives

$$\frac{\overline{DP}}{\overline{DP}_0} = (1 - C)^{(B-1)/B} \quad . \tag{6.26}$$

For random bond break depolymerization with partial unzipping, $\overline{DP}$ is a complicated logarithmic function of time (reaction extent), whereas for complete unzipping, $1/\overline{DP}$ is a linear function with time.

For chain-end initiation followed by partial unzipping:

$$-\frac{dN}{dt} = 0 \tag{6.27}$$

and from Eqs. (6.18), (6.21) and (6.27) with integration:

$$\frac{\overline{DP}}{\overline{DP}_0} = 1 - C \quad . \tag{6.28}$$

For chain-end initiation followed by complete unzipping:

$$-\frac{dN}{dt} = kN \tag{6.29}$$

and combining Eqs. (6.17), (6.18), (6.21) and (6.29) followed by integration yields

$$\frac{\overline{DP}}{\overline{DP}_0} = 1 \quad . \tag{6.30}$$

For end-group depolymerization, $\overline{DP}$ is a linear function of time for partial unzipping and independent of time for complete unzipping.

All the above mechanisms of depolymerization can be plotted as shown in Fig. 6.3 [130]:

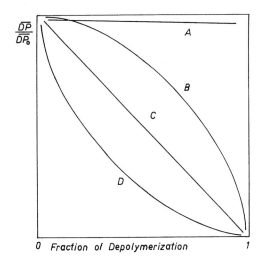

Fig. 6.3. Changing of average degrees of polymerization ($\overline{DP}$) as a function of fraction of depolymerization for different depolymerization (A-D) mechanisms (see text)

(i)   line A corresponds to initial end-group depolymerization followed by complete unzipping and shows no change in molecular weight since the initiation of depolymerization is the rate-depending step, i.e. chain group of remaining chains remains unchanged;

(ii) line B corresponds to random initial depolymerization followed by complete unzipping, resulting in a relatively slow decrease in molecular weight with conversion;

(iii) line C corresponds to chain-end initiation followed by partial (incomplete) unzipping, where average chain length will decrease slowly, the change in average chain length being related to the size of segment removed;

(iv) line D corresponds to random initial depolymerization followed by partial (incomplete) unzipping, resulting in a rapid decrease in molecular weight with conversion.

Lines B and D are members of a number of similar curves, their exact shapes being dependent on the exact (photo)depolymerization parameters. These curves can be computer-generated for varying situations.

Depolymerization may occur via more than one pathway. The above approach can be extended to cover these cases and others. Thus there exist several ways to determine the mode of depolymerization, with most requiring the products to be soluble and permitting molecular weight to be determined.

6.5 Kinetics of Photodegradation

The "rate of photodegradation" may be expressed by the rate of photolysis of a given chromophoric group, which yields free radicals:

$$-\frac{d[c]}{dt} = \phi I_0 (1 - e^{\varepsilon l[c]}) \tag{6.31}$$

where [c] is the concentration of a given chromophoric group, ϕ is the quantum yield of photolysis of this group, I_0 is the incident UV(visible) radiation intensity, ε is the extinction coefficient of the chromophoric group, and l is the effective path of the radiation in the reaction vessel (for polymers in solutions) or film thickness.

If the concentration of a given chromophore group in a polymer is small (trace of impurities), the rate of photolysis is also small, and Eq. (6.31) can be replaced by [361, 376, 795]

$$-\frac{d[c]}{dt} = \phi I_0 (1 - e^{\varepsilon l[c]_0}) \tag{6.32}$$

where $[c]_0$ is the initial concentration of chromophoric group and, since the extent of UV (visible) absorption is small (ε is small) and $[c]_0$ is small ($c_0 < 1$ per chain), Eq. (6.32) may be further simplified to

$$-\frac{d[c]}{dt} = \phi I_0 \varepsilon l [c]_0 \ .$$

(6.33)

When $[c]_0$ is small, Eq. (6.33) may be integrated (treating $[c]_0$ as constant) to give

$$[c]_0 - [c]_t = \phi I_0 \varepsilon l [c]_0 t \ .$$

(6.34)

The number of chain scissions per molecule (S) is related to the left hand side of Eq. (6.34) as

$$S = \frac{[c]_0 - [c]_t}{[X]_0}$$

(6.35)

where $[X]_0$ is the initial concentration of polymer chains [376].
    Substituting Eq. (6.34) in Eq. (6.35):

$$S = \frac{\phi I_0 \varepsilon l [c]_0 t}{[X]_0} \ .$$

(6.36)

For random chain scission kinetics as applied to small extents of degradation [361, 376], the degree of degradation (β) is given by

$$\beta = \frac{1}{\overline{DP}_t} - \frac{1}{\overline{DP}_0}$$

(6.37)

where $\overline{DP}_0$ and $\overline{DP}_t$ are the initial average degree of polymerization and after a period (t) of reaction, related to the average molecular weight of the polymer ($\overline{M}$) as follows:

$$DP = \overline{M}/m$$

(6.38)

where m is the molecular weight of the monomeric unit.
    Since

$$\beta = \frac{S}{\overline{DP}_0}$$

(6.39)

Eq. (6.39) becomes

$$\beta = \frac{\phi I_0 \varepsilon l [c]_0 t}{[X]_0 \overline{DP}_0}$$

(6.40)

which predicts a linearity of β with the time of photodegradation, as shown in Fig. 6.4.
    A detailed treatment of the kinetics of photodegradation of terpolymers has already been given elsewhere [489].

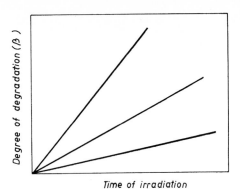

Fig. 6.4. Changing of degree of degradation (β) as a function of time of irradiation

6.6   Number of Main Chain Scissions

The "number of main chain scissions" (S) per original molecule is related to the initial $(\overline{M}_n)_0$ (for undegraded polymer), and to that observed after time t, $(\overline{M}_n)_t$ by the equation [400]

$$S = \frac{w}{\overline{M}_{n(0)}} \left(\frac{(\overline{M}_n)_0}{(\overline{M}_n)_t} - 1 \right) \tag{6.41}$$

where $\overline{M}_n$ is the number-average molecular weight, and w is the weight in grams of the irradiated polymer.

At random chain scission, the value of S is also related to the initial $(\overline{M}_v)_0$ (for undegraded polymer), and to that observed after time t, $(\overline{M}_v)_t$ by the equation [43, 509]

$$S = \frac{2w}{(\overline{M}_v)_0} \left(\frac{(\overline{M}_v)_0}{(\overline{M}_v)_t} - 1 \right) \tag{6.42}$$

where $\overline{M}_v$ is the viscosity-average molecular weight, given by the equation

$$[\eta] = KM_v^\alpha \tag{6.43}$$

where $[\eta]$ is the intrinsic viscosity (the limiting viscosity number, and K and α are constants available in the literature [421].

In this way, the "number of main chain scissions" (S) can be easily determined from the viscometric measurements according to the equation

$$S = \left(\frac{[\eta]_0}{[\eta]_t} \right)^{1/\alpha} - 1 \tag{6.44}$$

where $[\eta]_0$ and $[\eta]_t$ are the intrinsic viscosity of polymer before and after irradiation, respectively.

The linearity of the plot (S) vs (t) in the early stages of the degradation (Fig. 6.5) implies that random chain scission is occurring [164]. That this is not

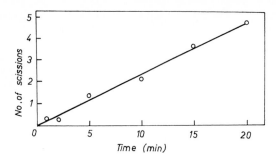

Fig. 6.5. Number of scissions per molecule of poly(phenylvinyl ketone) in benzene solution against irradiation time (at 365 nm) [164, 279]

maintained, however, suggests that other competing reactions, such as crosslinking, play a more significant role as the reaction proceeds.

6.7   Quantum Yield of Chain Scission

The "quantum yield for the main chain scission" (ϕ_s) (defined by Eq. 4.2) is given by

$$\phi_s = \frac{1}{(\overline{M}_n)_0} \frac{S}{E_s} \tag{6.45}$$

where E_s is the energy of radiation absorbed for the main chain scission (the scission dose) (in Einsteins per gram of polymer) or

$$\phi_s = \frac{w}{(\overline{M}_n)_0} \frac{S}{I_a t} \tag{6.46}$$

where w is the weight in grams of the irradiated polymer, $(\overline{M}_n)_0$ is the number-average molecular weight, S is the number of main chain scissions, I_a is the amount of radiation absorbed, and t is the time of irradiation.

Using data from the linear portions of the S vs t plot, it is easy to calculate ϕ_s. In Table 6.1 examples of quantum yields of chain scissions for different polymers are given.

Three main factors contribute to the low quantum yield for chain scission [792, 793].

(i)   Cage effects are more pronounced in the solid state, and hence the diffusive separation of two fragments is impeded. Because of the closer proximity of the molecules in the solid state, there is a higher probability that an abstractable H-atom will be sufficiently close to an alkoxy radical. This will enhance the competition from abstraction, which is already more favourable on account of the higher rate constant, i.e. typical rate constant for H-abstraction and decomposition are 10^6 mol s^{-1} and 10^4 s^{-1}, respectively.

(ii)   Crosslinking rapidly overwhelms the effects of scission. Crosslinking, which is facilitated by the proximity of the molecules, may be brought about by the interactions of alkoxy and peroxy radicals, and by others

Table 6.1. Quantum yields of chain scission

Polymer	Wavelength used (nm)	Quantum yield	References
Poly(methyl vinyl ketone)	253.7	2.5×10^{-2}	321
Poly(methyl isopropenyl ketone)	253.7	2.2×10^{-1}	696
Poly(phenyl isopropenyl ketone)	253.7	1.7×10^{-1}	506
	313.0	5×10^{-2}	506
Poly(methyl methacrylate)	253.7	$1.7\text{-}3 \times 10^{-2}$	247
Poly(vinyl acetate) in vacuum)	253.7	6.68×10^{-2}	162
Poly(vinyl acetate) (in air)	253.7	5×10^{-3}	774
Polystyrene (in benzene)	253.7	3×10^{-5}	581
Polystyrene (film)	253.7	5.5×10^{-4}	280
Poly(α-methyl styrene) (film)	253.7	7×10^{-3}	734
Polyacrylonitrile	253.7	$2.0\text{-}7.7 \times 10^{-4}$	373
Poly(ethylene terephthalate	253.7	1.6×10^{-3}	555
	280.0	5×10^{-4}	475
Polysulphones (film)	253.7	8.4×10^{-4}	422
Polyethersulphones (film)	253.7	7.7×10^{-4}	422
Poly(methyl phenyl silane) (in solution)	313.0	9.7×10^{-1}	764
Poly(methyl phenyl silane) (in film)	313.0	1.7×10^{-2}	764

which are formed on the chains as a result of abstraction by the reactive hydroxy (HO·) radical.

(iii) Decomposition of the hydroperoxide (POOH) by interaction with the carbonyl groups, particularly that involving energy transfer, requires a very specific geometrical orientation of the two species. As the flexibility of the system is diminished by crosslinking, it becomes more difficult to achieve these configurations, and hence the rate of POOH decomposition will decrease, and with it the rate of carbonyl groups formation.

6.8   Quantum Yield of Crosslinking

"Crosslinking" is the formation of new intermolecular bonds resulting in the binding together of macromolecules (Fig. 6.6). As the crosslinking proceeds, the molecular weight steadily increases until an insoluble gel is formed.

It has been shown [135] that no matter what the initial molecular weight distribution, three-dimensional network formation first begins to occur at a crosslinking density corresponding to one crosslinked unit per weight average molecule.

The proportion of monomer units (in a chain) crosslinked per unit radiation dose is given by

$$E_c = \frac{m}{M_w q_0} = \frac{1}{DP q_0} \tag{6.47}$$

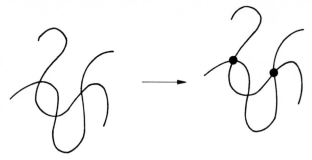

Fig. 6.6. Schematic presentation of crosslinking formation

where E_c is the energy of radiation absorbed for the crosslinking in the absence of the main chain scission (the gelling dose) (in Einstein per gram of polymer), m is the molecular weight of the monomeric unit, $\overline{M}_w$ is the weight average molecular weight, q_0 is the number of monomeric units (in a chain) crosslinked per unit radiation absorbed, and $\overline{DP}$ is the average degree of polymerization:

$$\overline{DP} = \frac{\overline{M}_w}{m} \ . \tag{6.48}$$

The "quantum yield for the crosslinking" (ϕ_c) is given by

$$\phi_c = \frac{\text{number of crosslinks}}{\text{number of quanta absorbed}} \tag{6.49}$$

and can be simply related to the energy of radiation absorbed (dose) for the crosslinking (E_c), recognizing that there are two crosslinked monomer units involved in each crosslink between polymer chains, by the equation

$$\phi_c = \frac{1}{2E_c} = \frac{1}{2}\overline{DP}q_0 \ . \tag{6.50}$$

During the photodegradation process, the crosslinking is almost always accompanied by the main chain scission (to some extent). In such a case the determination of quantum yields requires the measure of the soluble fraction (s) as a function of absorbed energy [135, 160, 162, 279]:

$$s + \sqrt{s} = \frac{p_0}{q_0} + \frac{2E_c}{E_s} \tag{6.51}$$

where p_0 is the number of chain scissions per unit of radiation absorbed, p_0/q_0 is the ratio of degradation to crosslinking, and E_s is the energy of radiation absorbed for the main chain scission (the scission dose) (in Einstein per gram of polymer).

In this case, the quantum yield for the crosslinking ($\phi_{c(s)}$) has to be calculated from E_c and the quantum yield of the chain scission (ϕ_s), is further determined from

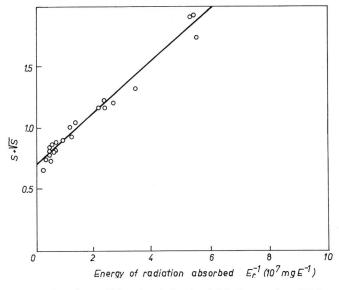

Fig. 6.7. Plot of $s + \sqrt{s}$ for photolysis of poly(vinyl acetate) at 253.7 nm vs inverse energy of radiation absorbed (E_r) [162, 279]

$$\frac{p_0}{q_0} = \frac{\phi_s}{2\phi_{c(s)}} \tag{6.52}$$

A typical plot of $s + \sqrt{s}$ vs reciprocal radiation energy absorbed is shown in Fig. 6.7.

Three limiting cases can be discussed (Fig. 6.8) [171].

(i)   If $p_0/q_0 = 2$, the probabilities for crosslink formation and chain scission are equal and $\overline{M}_w$ does not change with radiation energy absorbed.

(ii) If $p_0/q_0 < 2$, crosslinking is predominant. A three-dimensional network is formed at the energy of radiation absorbed for crosslinking (E_c) corresponding to $1/\overline{M}_w = 0$.

(iii) If $p_0/q_0 > 2$, main-chain scission is the determining factor and $\overline{M}_w$ decreases.

If a gel is formed, p_0/q_0 and the energy of radiation absorbed for crosslinking (E_c) can be obtained from solubility measurements. Experimental relation is obtained if the initial molecular weight distribution is random. The E_c and p_0/q_0 are obtained from the slope and intercept of the line, respectively, and q_0 can be calculated from Eq. (6.50) if E_c and $\overline{DP}$ are known. Then p_0 is determined from this q_0 value and the experimental value of p_0/q_0 (Fig. 6.9).

The "crosslinking coefficient" (δ) is defined as the number of units crosslinked per weight-average molecule and can be determined from the amount of soluble material (sol) (s) using the equation

$$s + \sqrt{s} = 2/\delta \;. \tag{6.53}$$

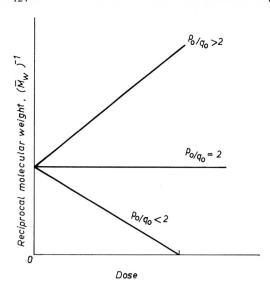

Fig. 6.8. Reciprocal weight average of the molecular weight ($\overline{M}_W$) as a function of the absorbed dose for different ratios of degradation to crosslinking (p_0/q_0) [135,160]

If the crosslinking process causes the decreasing of absorbance (A) (of a given chromophoric group) of the film, this can be related to the "quantum yield of chromophore removal" [648]:

$$\frac{dA}{dt} = I_a(\varepsilon_0 - \varepsilon_t)\left(\frac{A_0 - A_t}{A_0}\right)\phi \times 10^3 \qquad (6.54)$$

where I_a is the intensity of radiation absorbed (Einstein cm^{-2}s^{-1}), ε_0 and ε_t are extinction coefficients of a given chromophoric group before and after irradiation (during time t), respectively, A_0 and A_t are absorbances of a given chromophore before and after irradiation, respectively, and ϕ is the quantum yield of chromophore removal.

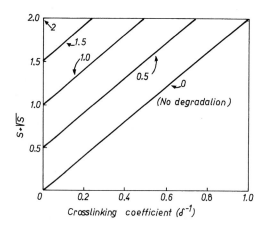

Fig. 6.9. Simultaneous crosslinking and degradation. Figures denote ratio of degradation to crosslinking (p_0/q_0) for an initially random distribution [135, 160]

6.9 Principles Governing the Changes
in Molecular Weight and Molecular Weight Distribution
During Polymer Photodegradation

Polymer photodegradation involves two processes: chain scission and/or crosslinking which change molecular weights ($\overline{M}_n$ and $\overline{M}_w$) and molecular weight distribution ($\overline{M}_w/\overline{M}_n$) [135, 161, 171].

In the absence of crosslinking, main chain scissions result in a decrease of both $\overline{M}_n$ and $\overline{M}_w$. For random scission $\overline{M}_w/\overline{M}_n$ tends to 2.

After 4–5 random scissions, the molecular weight distribution of a polymer becomes random ($\overline{M}_w/\overline{M}_n = 2$).

In the absence of chain scission (the crosslinking is the only process), $\overline{M}_w$ increases much more rapidly than $\overline{M}_n$ and has an infinite value at the gel point.

During simultaneous chain scission and crosslinking reactions, two cases must be considered [171].

(i)   If the probability of main chain scission is larger than the probability of crosslinking, $\overline{M}_n$ and $\overline{M}_w/\overline{M}_n$ decrease. The polymer remains soluble whatever the time of irradiation.

(ii) If the probability of chain scission is smaller than the probability of crosslinking, network formation is the determining factor. $\overline{M}_w/\overline{M}_n$ increases and an insoluble fraction of a gel appears. Complete insolubilization of the polymer is only observed if a main scission process does not occur.

An initially random distribution is described by the two equations

$$\frac{1}{(\overline{M}_n)_t} = \frac{1}{(\overline{M}_n)_0} + \left(p_0 - \frac{q_0}{2}\right)\frac{E}{m} \tag{6.55}$$

$$\frac{1}{(\overline{M}_w)_t} = \frac{1}{(\overline{M}_w)_0} + \left(\frac{p_0}{2} - q_0\right)\frac{E}{m} \tag{6.56}$$

where $(\overline{M}_n, \overline{M}_w)_0$ and $(\overline{M}_n, \overline{M}_w)_t$ are the average molecular weights of the polymer before and after a certain time (t) of irradiation, respectively, p_0 is the fraction of broken units in the main chain per unit incident energy, q_0 is the fraction of crosslinked units per unit incident energy (one crosslink involves two crosslinking units), m is the molecular weight of the monomer unit, and E is the total incident energy absorbed ($E_s + E_c$).

From the above equations it can be concluded (Fig. 6.10) [171] that

(i) if p_0/q_0 $\begin{cases} > 1/2 & \overline{M}_n \text{ decreases} \\ \\ < 1/2 & \overline{M}_n \text{ increases;} \end{cases}$

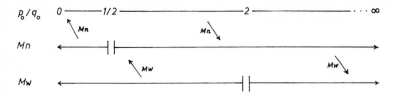

Fig. 6.10. Changing of degradation to crosslinking ratio (p_0/q_0) for different number ($\overline{M}_n$) and weight ($\overline{M}_w$) average molecular weights [171]

(ii) if p_0/q_0 $\begin{cases} > 2 & \overline{M}_w \text{ decreases} \\ < 2 & \overline{M}_w \text{ increases;} \end{cases}$

(iii) if $1/2 < p_0/q_0 < 2$ $\overline{M}_n$ decreases, $\overline{M}_w$ increases, and the molecular weight distribution broadens;

(iv) if $p_0/q_0 > 2$, the molecular weight distribution can either narrow or broaden according to the relative decrease of $\overline{M}_n$ and $\overline{M}_w$;

(v) if $p_0/q_0 \gg 2$, a random distribution characterized by $\overline{M}_w/\overline{M}_n = 2$ will be obtained after a small number of scissions per chain;

(vi) if $p_0/q_0 < 2$, $\overline{M}_w$ increases up to gel point where it has an infinite value. As a consequence, the distribution broadens as the number of crosslinks per chain scission increases.

The "number of main chain scissions" (S) per original molecule can be evaluated from the transformed Eq. (6.55):

$$S = \frac{\overline{M}_{n(0)}}{\overline{M}_{n(t)}} - 1 = \left(1 - \frac{q_0}{2p_0}\right) \frac{\overline{M}_{n(0)}}{m} p_0 \, E \ . \tag{6.57}$$

Gel permeation chromatography (GPC) is the simplest and most effective manner of monitoring the changes in molecular weight distribution occurring during photo (oxidative) degradation or natural (artificial) weathering or any similar degradation process [517, 541]. However, the technique has inherent shortcomings with respect to the level of reproducibility.

6.10 Reaction Kinetics of Polymer Photo-Oxidation

The accepted mechanisms for photo- and thermal oxidation of polymers are similar in many respects. The reaction kinetics of polymer photo-oxidation can be derived by considering the mechanistic steps of polymer photo-oxidation (cf. Sect. 4.9):

Initiation	PH $\xrightarrow{h\nu}$ P· + H·	k_1	(6.58)
Propagation	P· + O$_2$ $\longrightarrow$ POO·	k_2	(6.59)
	POO· + PH $\longrightarrow$ POOH + P·	k_3	(6.60)
Termination	2P· $\longrightarrow$ inactive products	k_4	(6.61)
	2POO· $\longrightarrow$ inactive products	k_5	(6.62)

(Note: in this scheme chain branching reactions were not taken into consideration.)

The reaction of O_2 with polymer alkyl radical (P·) is very fast and reaction of hydrogen abstraction by polymer peroxy radical (POO·) comparatively slow; typically $k_2/k_3 \gg 10^6$.

Under the steady-state conditions:

$$r_i = r_p = r_t \qquad (6.63)$$

$$r_i = k_1 I_a \qquad (6.64)$$

$$r_p = k_2[P·][O_2] + k_3[POO·][PH] \qquad (6.65)$$

$$r_t = k_4[P·]^2 \text{ at low oxygen concentration} \qquad (6.66)$$

$$r_t = k_5[POO·]^2 \text{ at high oxygen concentration} \qquad (6.67)$$

where r_i, r_p and r_t are the rates of initiation, propagation and termination, respectively, and I_a is the intensity of radiation absorbed by the polymer during photo-oxidation.

It is appropriate to note here that stationary state equations should not be applied indiscriminately to polymeric systems in which radical concentrations are likely to change rapidly at some stage, e.g. termination in the presence of radical scavengers (photoinitiators, thermostabilizers, radical traps, etc).

Photo-oxidation can be carried out under two experimental conditions [53, 81, 599].

(i)   Low concentration of oxygen, where termination occurs almost exclusively by the recombination of two polymer alkyl radicals (P·) (Reaction 6.61). Under these conditions, the concentration of polymer alkyl radicals (P·) is greater than that of polymer peroxyl radicals (POO·). In this case the rate of oxygen uptake or polymer hydroperoxide (POOH) formation is given by

$$-\frac{d[O_2]}{dt} = \frac{d[POOH]}{dt} = k_2[P·][O_2] \ . \qquad (6.68)$$

Thus assuming that

$$r_t = k_4[P·]^2 \qquad (6.69)$$

we obtain:

$$P\cdot = \left(\frac{r_t}{k_4}\right)^{1/2} .$$

(6.70)

Under steady-state conditions:

$$r_t = r_i = k_1 I_a .$$

(6.71)

Substitution of Eq. (6.70) into Eq. (6.69) gives, for the rate of oxygen uptake:

$$-\frac{d[O_2]}{dt} = k_2\left(\frac{k_1 I_a}{k_4}\right)[O_2] .$$

(6.72)

Equation (6.72) shows that, under conditions of constant radiation intensity and low oxygen concentration, the rate of oxygen uptake is proportional to the concentration of available oxygen that is present at any given time. In this case oxygen is the limiting reagent and thus controls the extent of reaction.

(ii)   High concentration of oxygen (atmospheric pressure), where the termination occurs almost exclusively by the recombination of two polymer peroxyl radicals (POO·) (Reaction 6.62). It is also true that, under these conditions, the rate of combination of polymer alkyl radical (P·) with oxygen (Reaction 6.59) is relatively fast and so [P·] ≪ [POO·]. Under these conditions, the rate of oxygen uptake or polymer hydroperoxide (POOH) formation is given by

$$-\frac{d[O_2]}{dt} = \frac{d[POOH]}{dt} = k_3[PH][POO\cdot] .$$

(6.73)

Thus assuming that

$$r_t = k_5[POO\cdot]^2$$

(6.74)

we obtain

$$[POO\cdot] = \left(\frac{r_t}{k_5}\right)^{1/2} .$$

(6.75)

Under steady-state conditions:

$$r_t = r_i = k_1 I_a .$$

(6.76)

Substitution of Eq. (6.75) into Eq. (6.73) gives, for the rate of oxygen uptake:

$$-\frac{d[O_2]}{dt} = k_3[PH]\left(\frac{k_1 I_a}{k_5}\right)^{1/2} .$$

(6.77)

Equation (6.77) shows that, under conditions of constant radiation intensity and high oxygen concentration, the rate is proportional to the concentration of polymer (PH) and is independent of oxygen pressure. This result also suggests that the rate of oxygen uptake is proportional to the number of polymer alkyl radicals (P·) at any given time that are available for the oxidation.

Equations (6.72) and (6.77) show that oxygen uptake under steady-state conditions is a first order process. This process undergoes a transition as the oxygen concentration of the system changes. The rate of photo-oxidation of the polymer depends on the ratio of available oxygen to the number of polymer alkyl radicals (P·), which is determined by the partial pressure of oxygen in the gas phase.

The effects of the partial pressure of oxygen and temperature on the rate of photo-oxidation are of considerable importance in the study of polymer photodegradation [80].

A number of different types of oxygen uptake apparatus for studying the photo-oxidation of polymers have been described elsewhere [80, 158, 305]. Typical oxygen uptake curves are shown in Fig. 6.11. The oxygen uptake curves differ one from another in thermal- and photo-oxidation. In the latter there is no (or very little) induction period.

The early stage of the photo-oxidative degradation is first order with respect to the partial pressure of oxygen and is described by the expression [80, 81]

$$n(t) = x_i(1 - e^{-kt}) \qquad (6.78)$$

where $n(t)$ is the number of moles of oxygen that have reacted with the polymer at any given time t, x_i is the number of moles of polymer alkyl radicals (P·) that react with oxygen, and k is the first order rate constant.

The "initial quantum yield for oxygen uptake" under irradiation (ϕ_{O_2}) is a measure of the inherent photostability of a polymer and can be calculated from parameters x_i and k. In particular:

$$\phi_{O_2} \propto \frac{mx_i k}{A_s \lambda} \qquad (6.79)$$

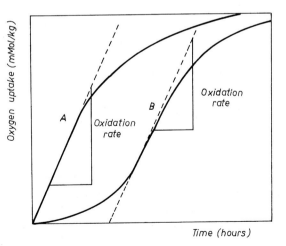

Fig. 6.11. Typical oxygen uptake curve (A = photo oxidation; B = thermal oxidation)

where m is the mass of the polymer sample (kg), and A_s is the surface area (m^2) of the sample exposed to the incident UV radiation of wavelength λ (nm).

An expression for the initial rate of photo-oxidation (r_{ox}) can be obtained by differentiating the function n(t) and evaluating it at time t $= 0$, in which case

$$r_{ox} = x_i k . \tag{6.80}$$

The values of x_i and k can be determined from the experimental photo-oxidation curve and its initial gradient, r_{ox}, by using a method of successive estimates of x_i [80]. If r(t) is the time derivative of the function n(t), then

$$r(t) = x_i k \, e^{-kt} . \tag{6.81}$$

Equation (6.82) expresses x_i in terms of the experimentally measurable parameters r_{ox}, r(t) and n(t). Successive estimates of x_i can be thus determined by the direct measurement of these parameters:

$$x_i = \frac{n(t)r_{ox}}{r_{ox} - r(t)} . \tag{6.82}$$

If x_i is known, a plot of $\ln[1-n(t)/x_i]$ vs time will be a straight line which passes through the origin and with a slope equal to k

A typical Arrhenius plot of the logarithm of photo-oxidation rate vs the reciprocal of the absolute temperature:

$$k = C \exp(-E_a/RT) \tag{6.83}$$

where C is a constant, E_a is the activation energy (kJ mol^{-1}), and T is the absolute temperature (K) can be used to determine activation energy (E_a) of the photo-oxidation process [65, 81, 303, 304, 496, 805].

The Arrhenius equation can be rewritten in terms of the initial rate of photo-oxidation (r_{ox}) by substituting $r_{ox} = x_i k$ in the equation, giving the expression

$$\ln(r_{ox}) = \frac{\ln(x_i C) - E_a}{RT} . \tag{6.84}$$

The above mentioned kinetics are valid only for very thin polymer films. However, in the case of a solid thick polymer sample, the situation may occur where the rate of reaction is determined by the availability of O_2 at the reaction site. In this case the diffusive behaviour of oxygen into the polymer sample plays an important role. A number of mathematical models have been proposed [83, 89, 229, 371, 802] to elucidate the behaviour of an oxidation reaction taking place in a thick polymer film, which take into account the diffusion behaviour of O_2.

Polymer photo-oxidation studies are often complicated by the fact that exposed samples are unhomogenously oxidized. At the macroscopic level the heterogeneities can result from oxygen-diffusion limited effects [155, 294]. If the rate of oxygen consumption exceeds the rate of oxygen permeation, oxidation occurs in the surface layers whereas the core remains practically unoxidized [196, 380, 381].The importance of this effect depends on several intrinsic parameters linked to material geometry. (e.g. sample thickness), coupled with the oxygen consumption rate, which depends on the reactivity of the polymer, the presence of additives and the oxygen permeability of the material. External parameters are the conditions of photo-oxidation of a polymer (accelerated ageing) and the oxygen pressure during photo-oxidation.

6.11   Diffusion Controlled Oxidation of Polymers

Mathematical models of diffusion controlled oxidation of polymers are generally based on the following assumptions [53, 229].

(i)   The rate of O_2 absorption is diffusion-controlled from the film surface, but in the case of thin film the rate is linearly proportional to the film thickness. The rate in this latter case is thus considered to be independent of diffusion effects.

(ii) The diffusion effects in a polymer film are generally expressed by a differential equation according to Fick's laws of diffusion. For any point in the film "Fick's equation" is

$$\frac{dC(x,t)}{dt} = D\frac{d^2C(x,t)}{dx^2} - kC(x,t) \tag{6.85}$$

where $C(x,t)$ is the concentration of O_2 at a given time t and given depth x assuming a steady-state condition is reached and maintained after a short reaction time, i.e. $dC/dt = 0$. k is the first-order rate constant
Equation (6.85) is only valid at the following assumptions:

(a)   rate of O_2 consumption proceeds at a rate dependent on and proportional to its concentration in the film, obeying first order kinetics;

(b)   the diffusion coefficient (D) of O_2 in the film is independent of O_2 concentration.

(iii) Oxygen diffuses into a solid polymer film through both surfaces and is consumed by a first-order or pseudo-first-order chemical reaction.

Diffusion controlled oxidation of polymers depends on a film thickness (Fig. 6.12).

(i)   Thin film of thickness L – in this case diffusion of O_2 is not a limiting factor on the oxidation reaction.

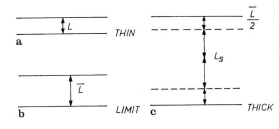

Fig. 6.12a–c. Schematic representation of the definition of limiting thickness [229]

(ii) Limit film of thickness $\overline{L}$ (limiting thickness) – in this case $\overline{L}$ is the maximum thickness of a sample in which diffusion effects of O_2 are negligible.

(iii) Thick film reached a value greater than $\overline{L}$ – in a centrum of such films exists a volume where diffusion of oxygen to reactive sites is the rate-determining step of the oxidation reaction. In such a thick film the diffusion-controlled zone exists extending from the bottom of the diffusion-free zone ($\overline{L}/2$) to the centre of the film (assigned as L_s).

Solution of Eq. (6.85), assuming the boundary conditions

$C = C_0$ at $x = 0$ and $t \geq 0$

$C = C_0$ at all x and $t = 0$

$C = 0$ at $x = \infty$ and $t > 0$

after final simplification gives

$$C(x, t) = C_0 \exp(-x\sqrt{k/D}) \ . \tag{6.86}$$

Figure 6.13 illustrates graphically Eq. (6.86) and the concentration gradient in the diffusion-controlled region of an oxidising polymer film. Combining this with the known concentration present in the diffusion-free outer regions of a

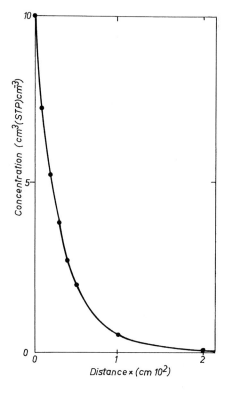

Fig. 6.13. Approximate solution of Eq. (6.86) for $C(x,t)$; variation of oxygen concentration with depth into film [229]

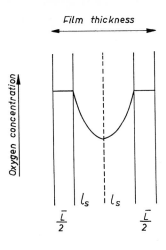

Fig. 6.14. Schematic representation of oxygen concentration in a degrading polymer film [229]

sample, it is possible to construct the behaviour of O_2 concentration within an oxidising sample (Fig. 6.14). The minimum value in the profile can reach zero at relatively low values of L_s.

6.12 Oxidation Profiles

Detailed information on the photo-oxidation of a polymer sample (film) can be obtained from the oxidation profiles as a function of the cross-sectional position (Fig. 6.15) [82, 83, 155, 196, 258, 351, 355].

The typical oxidation profile depends on:

(i)   sample thickness;
(ii)   oxygen diffusion and solubility in a polymer matrix;
(iii) oxygen pressure;
(iv)   mode of sample irradiation (one- or two-sided exposure);
(v)   percentage of crystalline phase (in semicrystalline polymers);
(vi)   tensile stresses;
(vii) the presence of cracks, pinholes and voids leading to changes of oxidation profile.

Oxidation profiles are symmetrical with respect to the centre of the sample stack (Fig. 6.15); however, photoirradiation from one side of a film makes them unsymmetrical (heterogenous oxidation).

Two methods are generally used for the determination of the oxidation profiles [196].

(i)   The intensity of an oxidation band can be monitored as a function of the sample thickness [267]. This method does not allow an oxygen-diffusion effect to be distinguished from a limitation of the degradation by UV absorption either by an initial chromophore or by UV absorbing photoproducts.

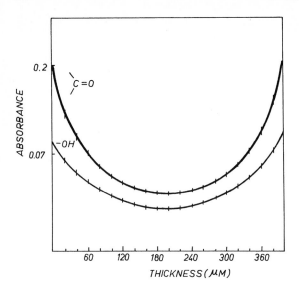

Fig. 6.15. Oxidation profiles as a function of the cross-sectional position [702]

(ii) The intensity of an oxidation can be monitored for microtomed slices as a function of the cross-sectional position [25, 665]. This method is not easily applied to the analysis of thin films and is usually reserved for monitoring heterogenous oxidation of thick films (a few millimeters in thickness). Moreover the brittleness of the exposed outer layers makes it difficult to collect the microtome cuttings. Consequently the samples often consist of a powder which must be mixed with KBr and pressed into small pellets. The data collected from the analysis of the first layers near the surface are often less accurate in such conditions.

Another method that is very sensitive to surface oxidation is attenuated total reflection (ATR) spectroscopy [601, 608, 609]. This method can be applied to both thick and thin films. By varying the reflecting elements and the angle of incidence it is possible to monitor the depth dependance of oxidation in the range of 0.1 μm to few micrometers [123, 196, 458]. However, this technique requires intimate contact between the polymer sample and the reflecting crystal.

This contact is not always possible to obtain with degraded materials since the outer surface roughness and often becomes brittle and deformed. Poor quantitative data are collected most of the time. Roughened surfaces also result in reduced spectral quality.

Much more sophisticated is micro-FTIR spectroscopy, which allows ana-lysing of the photo-oxidized sample in a plane perpendicular to the axis of irradiation. For this purpose, the irradiated film is first imbedded in a resin and then sliced with a microtome in a plane corresponding to the irradiation axis (Fig. 6.16) [196, 259, 381]. The thin slices (typically 50–80 μm) are then analysed by transmission IR spectroscopy through an IR microscope. A small area is isolated by an image-masking aperture and the oxidation profile is

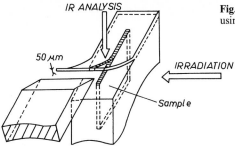

Fig. 6.16. Principle for analyzing thin films using micro(FTIR)-spectrophotometry [259]

monitored by moving the analysed area from the front layers to the back layers. This method has been successfully applied to the characterization of many different polymer systems.

The newest method for the study of photo-oxidized samples is photo-acoustic-FTIR spectroscopy, which does not require sample preparation and is relatively independent of the surface morphology of the sample [601, 609]. This method has been used for the study of photo-oxidation of polymer materials [196].

7　Photodegradation of Solid Polymers

7.1　Differences Between Photodegradation of Polymers in Solid and Liquid Phases

There are several reasons why the photodegradation processes in solid polymers differ from those proceeding in the liquid phase [561, 589, 769]. Differences between solution- and solid-photodegradation arise from:

(i)　mobility of macromolecules;
(ii)　diffusion-controlled reactions;
(iii)　solvent-macromolecule interactions;
(iv)　reactions caused by free radicals formed from the solvent, photolysis;
(v)　cage effect.

The free volume in polymeric matrix increases with temperature so that diffusion of oxygen and other small molecule reagents or free radicals becomes possible, even at temperatures below 100 K.

The diffusive motion of oxygen and/or small molecules (radicals) occurs by a series of hops as free volume is opened up by segment motion in spaces adjacent to the moving molecule without, however, net translational movements of the centre of mass of the polymer.

For amorphous polymers above the glass transition temperature (T_g), the rates of diffusion of small molecules are only one or two orders of magnitude below those in ordinary liquids. In Table 7.1 diffusion activation energies of oxygen and other gases in polymers [798] are shown. The relatively high mobility of small molecules in polymers above their glass transition temperature and the almost complete conformational freedom of the polymer chain means that reaction may take place under these conditions nearly as rapidly as in solution.

As a polymer chain is crosslinked the coordinated segment motion available for permeation is reduced, with a consequent reduction in the diffusion coefficient.

In solid polymers, the general stochastic nature of all degradation processes is likely to be more strongly expressed. Heterogeneity of every real solid polymer, fluctuation in local mechanical stresses, supermolecular structure and orientation [7, 429, 590] are important factors contributing to the heterogeneous nature and even anisotropy of degradation processes. This is especially pronounced in semi-crystalline polymers. Moreover, exposure of a solid polymer to natural weathering or accelerated ageing may create thermal

Table 7.1. Activation energy for diffusion of O_2, N_2 and H_2 in polymers [798]

Polymer	Activation energy for diffusion $(kJ\ mol)^{-1}$		
	O_2	N_2	H_2
Poly(tetrafluoroethylene	26.3	29.8	–
Perfluorinated ethylene-co-propylene	34.7	38.5	25.1
Polystyrene	37.7	–	16.7
Poly(ethyleneterephthalate)	50.6	58.7	–
Poly(N,N'-12-(p,p-oxydiphenylene)pyromellitide (Kapton)	–	31.0	–
Low density polyethylene (LDPE)	40.2	50.3	33.5
High density poly-ethylene (HDPE)	36.8	39.8	–
Polycarbonate	32.2	–	20.9
Polypropylene	36.5	41.9	34.7
Poly(methyl methacrylate)	–	20.9	–
Polyamide (Nylon 6)	33.5	46.1	31.5

and/or concentration gradients [258]. As a result, photodegradation in the solid state differs from that in the liquid state, not only in kinetic but also in mechanistic aspects. The degradation of a polymer in the solid state is a heterogeneous process, initiated and concentrated predominantly around impurities or weak centres.

Mechanical measurements are suitable and sensitive tools for following the processes during the photo-oxidative ageing of polymers in the solid state [8, 561, 589, 670]. Structural information can be obtained by combining mechanical data with structural imperfections [590]. Mechanical properties are a direct practical measure of material deterioration.

7.2 Polymer Mechanical Properties Affected by Degradation

Some of the most dramatic changes that occur when a polymer is UV(light) irradiated are in its mechanical properties; moreover, from an engineering point of view, these are perhaps of most importance.

The UV(light) degradation of thick polymer parts (>1 mm) is usually limited to a thin surface layer, which reaches a thickness of from a few hundred micrometers up to 1 mm at maximum [155, 229, 258].

Typical effects of degradation processes (chain scission and/or crosslinking) on physical properties of a degraded polymer are shown in Fig. 7.1 [474].

In general, an increase in tensile strength, modulus, hardness, density, etc. is interpreted as indicating a predominance of the crosslinking process over either or both of the chain scission processes. At the same time, such properties as elongation at break, viscoelastic flow (creep), etc. are expected to show a decrease in value with increased degree of crosslinking.

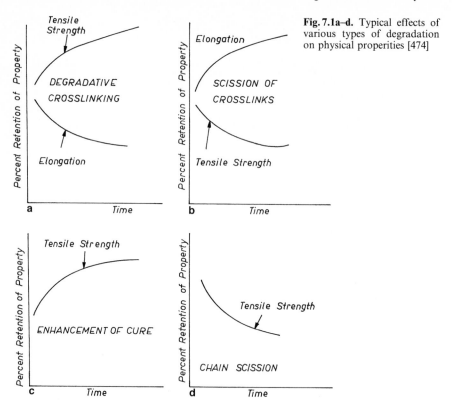

Fig. 7.1a–d. Typical effects of various types of degradation on physical properities [474]

The presence of mechanical stress during UV(light) irradiation may alter the relative amounts of crosslinking and scission, as well as having an effect on the spatial distribution of crosslinks that are formed.

Degradation processes often cause a transition from ductile failure to brittle fracture, thereby decreasing considerably the residual strength.

Fracture usually occurs from surface cracks (Fig. 7.2) which form easily in the degraded layer. The morphology (shape, length, depth) depends on type of polymer and conditions in which a given polymer is irradiated [76, 270, 663–665]. The residual strength of a cracked sample will depend on the crack length, i.e. the depth of embrittlement. Microcracks act as stress concentratators and fracture loci.

7.3   Role of Water in the Degradation Processes

Polymeric materials in the presence of water (high atmospheric humidities, dew, rain or snow) undergo several chemical and physical processes which seriously affect the mechanical properties of the material. As water permeates into the polymer matrix, it behaves as a plasticizer causing a diminution in

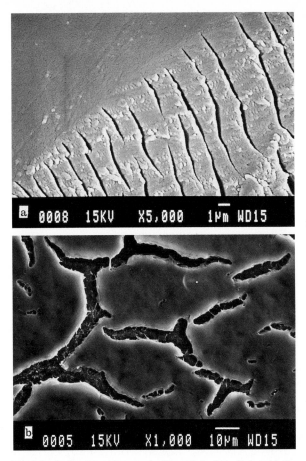

Fig. 7.2a, b. Surface microcracks formed during photooxidation of surfaces of: **a** poly(ethylene oxide) [382]; **b** poly(methacrylic acid) (provided by Dr H. Kaczmarek, Copernicus University, Torun, Poland)

tensile strength and often an increase in ultimate elongation. The compressive and tensile stresses which occur as a result of water absorption and desorption for a plate of thickness (d) are shown in Fig. 7.3. During water absorption, compressive stresses dominate externally while tensile stresses dominate in the interior; correspondingly, during water desorption tensile stresses dominate externally and compressive stresses internally. These stresses disappear once moisture equilibrium is achieved [406].

Water absorbed from humid air and from direct wetting by rain or dew causes swelling of plastic components or of coatings. A subsequent dry period will then entail the loss of water. Drying out of the surface layers would lead to volume contraction, which is, however, hindered, since the underlaying layers are still swollen. This process gives rise to tensile stresses on the surface, which may result in cracking [406].

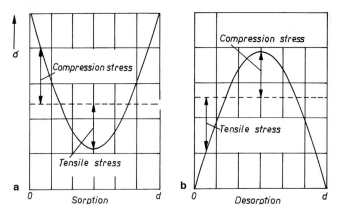

Fig. 7.3a, b. Tensile and compressive stresses; **a** during swelling (sorption); **b** during contraction (desorption) [406]

Water absorption is a diffusion-controlled process. The time determining factor is the diffusion coefficient, which is of the order of magnitude 10^{-8} cm^2 s^{-1} in plastics. This signifies that weeks or even months will pass before moisture equilibrium is achieved in compact plastics. The more rapidly humidity occurs at the surface, then in the deeper layers. The changes in humidity with time casue the mechanical stresses.

Plasticizers, processing aids and other additives present in plastics are leached out causing a decrease in elongation and an increased tensile strength. Migration of additives through a polymer matrix is accelerated by temperature. A buildup of plasticizers or stabilizers on the surface of the polymer is often manifested in appearance of a hazy (strongly sunlight absorbing) layer or "bloom". This layer can affect the polymer photodegradation in two different ways:

(i)   it absorbs UV radiation and protects the polymer against photodegradation;
(ii) to become photolysed, products of its photolysis then reacting with a polymer surface to initiate the oxidative degradation of a polymer.

7.4   Effect of Orientation on Degradation

The rate of polymer oxidation can be changed by changing the sample morphology by orientation drawing [73, 76, 111, 516, 687]. Both uniaxial and biaxial elongation enhance the photodegradation of low density polyethylene [76, 111, 516]. Highly oriented polymer material is less sensitive to photodegradation. The increase in degradation rate may be attributed primarily to strain effects (morphological changes) with some contribution from stress (stored energy). Biaxial stretching causes a greater effect on photodegradation, probably because of a larger decrease in film thickness and the greater constraint applied [133, 694].

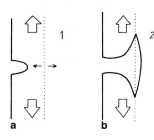

Fig. 7.4a, b. Formation of a microcrack in an anisotropic material: **a** a crack approaching a plane of weakness; **b** building up the main crack [591]

Degradation of oriented semicrystalline polymers causes an anisotropic embrittlement which is lower along the stretching direction and decreases with increasing orientation [429]. The fibrillation appears on the fracture surfaces if the specimen has been loaded along the orientation direction [591].

In an uniaxially loaded sample with a crack, there are two types of stresses around the crack tip:

(i)   stresses parallel to the external force show a maximum at the crack tip;
(ii) tensile stresses acting in the cross direction with a maximum located somewhat ahead of the crack tip.

In the elastic case the maximum of the cross stresses can be one-fifth of the longitudinal stress concentration. This causes a plane of weakness parallel to the external stress and perpendicular to the crack path to be opened. This can blunt and deflect the growing crack (Fig. 7.4). If cohesion perpendicular to the orientation direction is lowered under degradation, the toughness and extensibility of the material are less affected in the orientation direction. However, if the longitudinal cohesion drops below a certain critical value, the crack blunting mechanism is no longer effective and a dramatic embrittlement occurs even along the orientation. A loading of a polymer below the yield point during the photo-oxidative degradation can cause this type behaviour [591].

7.5   Degradation of Polymers Under Deformation Forces

In an environmental condition most practically utilized polymers and plastics work under different types of deformation forces (processes). These forces cause degradation of a polymer.

The mechanisms operative during mechanical deformations of amorphous polymers are both well documented and well understood through numerous thin-film and macro studies [412, 413]. Such models are described within a framework of the entanglement model with two possible routes envisaged for creation of the void-fibryl craze microstructure: chain disentanglement or chain scission.

It has been shown [204, 575] that the process of disentanglement, which involves the relative motion of one chain past another, is favoured by high temperatures, low molecular weights and low strain rates. On the other hand, chain scission involves the direct breaking of chemical bonds in the backbone, and leads to a substantial energy penalty, which increases in magnitude as the density of entanglement points is increased [412]. Thus scission crazing is less likely develop in highly entangled polymer networks. The transition to a more brittle response at high temperature for what is normally a tough polymer, is identified with a switch in deformation from shear yielding to disentanglement-mediated crazing [575]. Shear is always a potential competing mechanism with crazing, and the process that occurs for the lowest stress will develop first.

The nature of crazing in semicrystalline polymers, and the associated mechanisms, are much less well understood than in amorphous polymers. Part of the problem relates to the variability in microstructure in semicrystalline polymers and the subsequent complex interaction between morphology and deformation. Possible void-generating mechanisms involves a combination of interlamellar slip, chain disentanglement, block rotation and fragmentation processes [256, 507].

Using the methods of thermodynamics applied to irreversible processes offers a new approach for understanding the failure of polymers under deformation forces. For example, the equilibrium thermodynamics of closed systems predicts that a system will evolve in a manner that minimizes its energy (or maximizes its entropy). The thermodynamics of irreversible processes in open systems predicts that the system will evolve in a manner that minimizes the dissipation of energy under the constraint that a balance of power is maintained between the system and its environment. Application of these principles of nonlinear irreversible thermodynamics has made possible a formal relationship between thermodynamics, molecular and morphological structural parameters, rather than a simple sum of effects of individual stresses and their rates of change [132, 453].

Experimentally, these principles relate to dynamic measurements that make possible the separation of the dissipative and conservative components of energy incident upon the system. Dynamic mechanical analysis and computer-controlled experimentation now make it possible to apply analogous techniques to the measurement of many other thermodynamic stresses. For example, dynamic photothermal spectroscopy provides a new approach to predicting the long-term effects of UV radiation on materials [453].

7.6 Feedback in Polymer Degradation

The combined effect of chemical and physical environment and stress on a polymeric material can be described in terms of feedback, as the chemical and physical changes during exposure (output) affect in turn further progress of the degradation, deformation and fracture processes (Fig. 7.5) [591].

Changes of morphology and particularly of orientation act as a negative or inverse feedback, because orientational strengthening and plastically deformed zones at crack tips stop the progress of fracture and deformation [822].

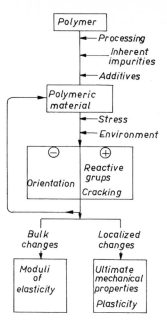

Fig. 7.5. Positive and negative feedbacks in degradation, deformation and fracture [591]

The orientation generally slows down the rate of degradation along the stretching direction [429, 590] as monitored by ultimate elongation. A system with a negative feed back can behave as an oscillation formed during crack or neck propagation.

On the other hand, cracking and formation of reactive groups cause a positive feedback and the system behaves as an amplifier. The stress concentration at the crack tip increases with increasing crack length. The amount of reactive groups formed during degradation and their subsequent transformations further accelerate the degradation process. A specific case of these feedback processes may involve chemical transformation of stabilizers and antioxidants.

7.7   Stress-Accelerated Photo-Oxidation of Polymers

There are few studies of stress accelerated photo-oxidation in the literature [66, 67, 539, 540]. The rate of photo-oxidation (of polystyrene) increases under applied stress [540]. A tensile stress increases the oxygen diffusion rate (in polypropylene film), and over an extended period of time it is possible that stress may cause an increase in crystallinity and this is expected to influence the rate of degradation [66]. The tensile stress accelerates molecular chain scission in injection-moulded bars (of polypropylene) exposed to ultraviolet radiation [539]. Stress may influence the rate of recombination of radicals formed in the chain scission process, effectively pulling them apart before the macroradical formed may recombine again, and/or it may influence the diffusion rate of oxygen to the radicals formed. Non-photochemical oxidation of stressed polymers has been discussed elsewhere [576].

7.8 Photodegradation of Polymers for Solar Energy Devices

During the last decade, considerable interest has developed in the application of polymers in the following areas of solar energy utilization [131, 132, 498, 605, 607]:

(i)   solar energy technologies;
(ii) solar energy storage;
(iii) photosensitized reduction of water in microheterogenous systems;
(iv) photoelectrochemical systems (photovoltaic cells, photoelectrochemical cells, etc);
(v)   artificial photosynthesis.

Polymeric materials are used in all solar energy technologies. In addition to such conventional applications as adhesives, coatings, moisture barriers, electrical and thermal insulation, and structural members, polymers are used as optical components in solar systems. Parabolic mirrors are made up of metallized fluoropolymers and acrylics. Commercial flat-plate collectors are glazed fluoropolymers and ultraviolet-stabilized polyester-glass fiber composites. Photovoltaic cell arrays are encapsulated with silicones and acrylics for protection from the weather. In laminated (safety glass) mirrors for central receiver heliostat systems, poly(vinyl butyral) is used as a laminating and encapsulating agent. Cast and moulded acrylic Frensel lenses are used to concentrate sunlight onto photovoltaic cells and thermal receivers. This widespread use of polymers in solar energy technologies is shown in Table 7.2.

Application of polymeric materials to solar energy devices offers a number of economic and performance advantages. Polymeric materials are lighter weight, easier to process and lower cost.

Table 7.2. Applications of polymers in solar energy conversion [132]

Solar Energy System	Mirror Glazings	Flat-Plate Glazings	Encapsulation	Seals/Adhesives	Structural Members	Heat Transfer/Energy Storage	Paints/Coatings	Piping	Thermal & Electrical Insulation	Moisture Barriers	Fresnel Lenses
Solar thermal conversion	×	×	×	×	×		×	×	×	×	×
Photovoltaics	×	×	×	×	×	×	×	×	×		×
Solar heating and cooling of buildings	×	×	×	×	×		×	×	×	×	×

In order to function properly in solar devices over many years of uses, polymeric materials must exhibit good long-term resistance to solar radiation, high temperatures, thermal and humidity cycles and mechanical stresses. The simulataneous action of UV radiation and high termperatures (even to 200°C) causes two parallel processes – (degradative) scission (fragmentation) and (degradative) crosslinking of chains. With increasing temperature, additives (low molecular weight compounds) such as plasticizers, processing aids, thermostabilizers, photostabilizers and antioxidants, diffuse at increasing rates out of polymer matrix. Loss of these additives causes the polymeric material to become more susceptible to photo-thermal ageing processes [608].

Distortion of a polymeric material can be generally caused by a combination of chemical reactions and mechano-physical processes. Under the simultaneous action of heat and pressure, the polymer chains rearrange to accommodate the applied forces. When the pressure is released, the newly-formed alignments undergo some structural reorganization, but nevertheless a complete recovery of the material to its original structure (and dimensions) occurs. Polyamides and polyurethanes and especially susceptible to exhibiting high compression changes because of dislocation of hydrogen bonds formed between macromolecules. When degradative processes (crosslinking and/or chain scission) occur to any appreciable extent, a polymeric material will exhibit very poor recovery or even none (i.e. permanent deformation). Toughness and ductility are important in allowing the polymeric material to sustain thermal expansion or mechanical torsion without brittle failure.

In prolonged exposure of polymeric materials in solar devices, other processes such as lost of transparency (increased absorption), colouration (yellow to brown and even black), chalking, delamination, and increased dirt retention may occur.

8 Photodecomposition of Polymers by Laser Radiation

8.1 Ablative Photodecomposition of Polymers

Pulsed ultraviolet laser radiation (at power densities greater than 1 MW cm^{-2} and a laser pulse of 1 μs width) from an excimer laser (producing wavelengths of 193, 248, 308 or 351 nm) causes etching of the polymer surface and the explosive ejection of the decomposition products at supersonic velocities (Fig. 8.1). This phenomenon has been termed ablative photodecomposition, and widely reviewed in several publications [57, 208, 432, 433, 709, 711, 712, 715, 719, 720, 726, 729].

Two mechanisms have been considered for the UV laser ablation of polymers [374, 529, 710, 717, 727, 728].

(i) Photochemical decomposition of the initially formed electronically excited states. In any polymeric system excited with photons of energy greater than about 3.6 eV ($\lambda < 340$ nm), the decomposition and ablation is predominantly photochemical, being caused by the excitation of chemical bonds to energy levels that are above the dissociation energy. The result is the scission of bonds and the production of a large number of small, volatile fragments during the absorption of the radiation pulse.

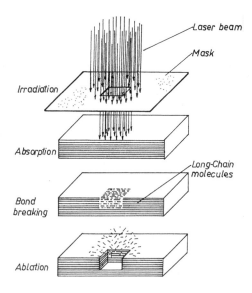

Fig. 8.1. Hypothetical steps in the interaction of a UV laser pulse with a polymer. *Top*: the laser radiation which is defined by a mask is absorbed. *Middle*: chemical bonds in the polymer are broken by the photon energy. *Bottom*: the products ablate at supersonic velocities leaving an etched sample [715]

Since the numerous fragments need a large free volume, an enormous pressure builds up in the small volume within the polymer at the site of irradiation, and they are ejected explosively [214]. Presumably the process is too quick for the fragments to transfer heat to the polymer and the excess energy is carried out. The etching is "clean", without charring or melting, substantially supporting a photochemical mechanism.

(ii) Photochemical decomposition in which the electronically excited state undergoes internal conversion to a vibrationally ("hot") excited ground state. Subsequent thermal decomposition is therefore a pyrolysis of the polymer, not very different from the processes that are observed with the laser radiation at visible or infrared wavelengths [46, 57, 100, 199, 411, 529]. This mechanism becomes increasingly significant at higher wavelength and/or fluence (cf. Sect. 8.3). Two types of thermal processes can occur, the photothermal process (or photopyrolysis) and the thermally activated photoprocess.

In the above two mechanisms, the formation of the electronically excited states of a polymer is a very essential step.

The photon flux from a pulsed UV laser interacts with a number of absorbing species (e.g. atoms, ions, molecules, etc.) in a polymer sample. These absorbing species are excited to the higher excited singlet and/or triplet states. The efficiencies of these transitions and lifetimes of the excited states are characteristic of the polymeric material and the laser photon energy. The higher excited states are generally harder to populate (due to a lower absorption cross section (cf. Sect. 1.3) and shorter lifetimes). This means that, for low intensity, the absorption of radiation only results in the depopulation of the ground state and therefore can be described by a single rate Lambert-Beer equation (Eq. 1.8). In this case, the absorption does not depend on the intensity, but decreases exponentially with the sample thickness. However, by using intense laser pulses, other energy levels are also populated. The absorption and the different population densities thus depend on the intensity. The Lambert-Beer equation is therefore invalid. Unfortunately, the time and spatial dependences of different population densities and the photon flux are unknown since only very few parameters (cross section and lifetime) can be measured experimentally (by using absorption and fluorescence spectroscopy). Because ablation occurs in the solid polymer matrix, the photoreactions are diffusion controlled ("cage reactions") and therefore multimolecular processes must be considered (e.g. annihilation (cf. Sect. 3.18) and collisional re-combination).

A detailed analysis of the mechanism of the UV laser ablation of a given polymer requires a knowledge of:

(i)   the energy expended in the material that is ablated;
(ii) an analysis of the products of the reaction;
(iii) the time in which ablation occurs.

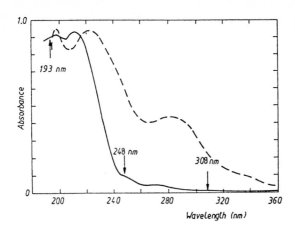

Fig. 8.2. Absorption spectra of:
(—) poly(methyl methacrylate)
and (- - -) polyimide [738]

The amount of energy absorbed by a polymer depends on its absorbance (A). In Fig. 8.2 absorbances of different polymers, i.e. poly(methyl methacrylate) and polyimide, are shown. Due to a strongly delocalized electronic distribution, polyamide presents absorption roughly two orders of magnitude higher than for poly(methyl methacrylate).

Polymers can be classified into two groups depending on their susceptibility towards ablative photodecomposition [529].

(i)   The low absorbing polymers (polyethylene, polypropylene, poly(vinylidene fluoride) where laser treatment does not produce ablation of polymer, although under high fluence and high repetition rate, melting occurs.

(ii) The low absorbing (poly(methyl methacrylate)) and high absorbing (polystyrene, polyimides, polysulphones, etc.) polymers where laser treatment produces ablation with a given threshold fluence. The stronger the absorption of UV radiation by a polymer, the lower the threshold.

Polymers with low absorbance can be efficiently ablated by incorporating suitable absorbing dopants (e.g. dyes) [50, 339, 483, 718, 723].

Introduction of acridine in concentrations of 1–8 wt% in poly(methyl methacrylate) progressively reduces the threshold for ablation at 248 nm and also sensitizes it to laser etching at 308 nm [723].

The etch depth per pulse (l_f) in a polymer surface caused by absorption of laser radiation exhibits a logarithmic dependence on the incident laser fluence (E_0) above a fluence threshold value (E_t) (Fig. 8.3) [46, 100, 212, 374, 717, 723, 724, 738]:

$$l_f = \frac{1}{A}\ln\frac{E}{E_0} = \frac{1}{\varepsilon M}\ln\frac{I_0}{I_t} \tag{8.1}$$

where A is the absorbance ($A = \varepsilon M$), ε is the extinction coefficient (also called "molar absorptivity" in $l\ mol^{-1}\ cm^{-1}$), M is the molar concentration (mol l^{-1}), I_0 is the intensity (fluence) of the incident laser beam, and I_t is the intensity threshold (fluence threshold) for the onset of ablative photodecomposition.

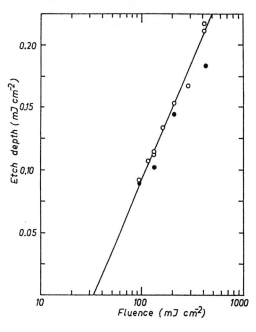

Fig. 8.3. Etch depth vs fluence in poly(ethylene terephthalate) with 193 nm laser radiation: (o) irradiation in vacuum; (•) irradiation in air [717]

The etch depth per pulse (l_f) cannot be derived from simple kinetics because the process of ablation represents a secondary, discontinuous phenomenon similar to explosion. I_t can be inserted into Eq. (8.1) from the knowledge that, when $l_f = 0$, $I_0 \neq 0$.

Equation (8.1) requires that a plot of l_f vs log I_0 has to be a straight line of slope $1/\varepsilon[M]$.

The etch depth per pulse at a given fluence is invariant with the depth of the etching (Fig. 8.4).

If the fluence is such that $l_f < l_0$, which is usually the case, then a depth $l_a - l_f$ (Fig. 8.5) which had been exposed to the radiation will be left behind.

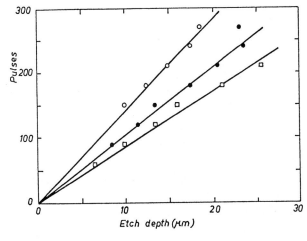

Fig. 8.4. Etch depth in polycarbonate as a function of number of pulses (193 nm). Laser fluence: (o) 165 mJ cm^{-2}, (•) 220 mJ cm^{-2}; (□) 300 mJ cm^{-2} [717]

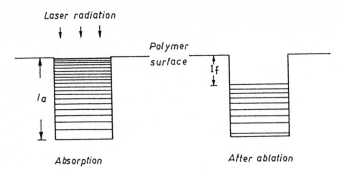

Laser radiation

Polymer surface

I_a

I_f

Absorption

After ablation

Fig. 8.5a, b. Schematic representation of the impact of laser pulse on a polymer surface [712]

The next pulse will go through this partly irradiated material as well as through virgin material underlying it. After the first pulses, there is a linear relationship between the number of pulses and the depth that is etched. In practice, the depths etched by varying numbers of pulses are averaged and noted as the etch depth per pulse for that polymer at the wavelength and fluence [473]. This value is reproducible within the uncertainty ($\pm 8\%$) in the measurement of the etch depth and the fluence of the laser pulse. This is the reason that the etch depth can be reproduced to ± 2000 Å in most polymers [712].

In the case of poly(methyl methacrylate), the first few photons reaching the polymer surface do not cause the etching process, but only melt the polymer

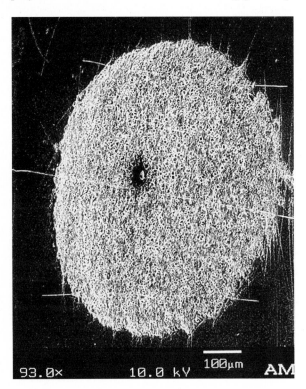

93.0×   10.0 kV   100μm   AM

Fig. 8.6. Scanning electron microscope photomicrograph of poly(methyl methacrylate) surface after 10 pulses at 248 nm [715] (provided by Dr R. Srinivasan, UVTech Associates, N.Y., USA)

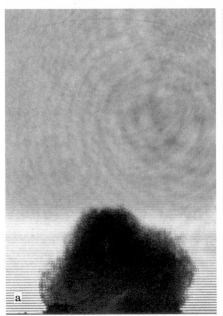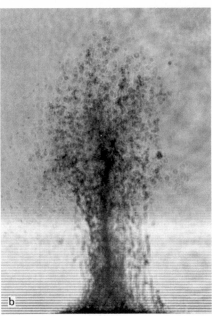

Fig. 8.7a, b. Sequence of photographs showing the development of the ejection of dense material during the ablation of poly(methyl methacrylate) by a single laser pulse (248 nm): **a** 6.1 μs; **b** 9.7 μs [715] (provided by Dr R. Srinivasan, UVTech Associates, N.Y., USA)

surface giving re-solidified material filled with gas bubbles (Fig. 8.6). This is the phenomenon of "incubation" (characteristic of poly(methyl methacrylate but not observed in the UV laser interaction of any other polymer) [715, 721]. However, the next UV laser pulses cause the development of the blast wave and the ejection of dense material (Fig. 8.7) and steadily etch the polymer surface.

The laser ablation of poly(methyl methacrylate) at 193 nm shows no temperature dependence, but at 248 nm reveals a strong temperature depend-ence [101].

The products of UV laser ablation range from atoms and diatoms to small polyatomic molecules and small fragments of the polymer [100, 712, 724, 727, 728, 816].

The products of laser ablation also depend on the molecular weight of UV ablated polymer.

(i) From poly(methyl methacrylate) of initial molecular weight $\overline{M}_n = 10^6$ at 193 nm, the products are carbon oxides, methyl methacrylate monomer, and low molecular weight ($\overline{M}_n = 1500$) fragments of the polymer. At 248 nm the principal product is a low molecular weight fraction ($\overline{M}_n = 2500$) of the polymer [723, 724].

(ii) From polyimide, the products are carbon oxides, benzene, HCN, and elemental carbon [722, 725]:

$$\begin{array}{l}
C_3H_2 \quad\quad CH \quad\quad\quad C_2H_2 \\
\quad CN \;\; C_2 \quad C_2 \quad CN \quad CO \\
\quad C_2H_2 \quad\quad CH \quad\quad\quad C_2H_2
\end{array} \qquad\qquad (8.2)$$

Molecular weight (e.g. polystyrene) affects photoablation [445].

The fragments of laser ablation are ejected (Fig. 8.8) with high velocity as large as $6–7\times10^5$ cm s^{-1} (Fig. 8.9).

As a result of the ablative photodecomposition of polymers by excimer lasers, different stable microstructures on polymer surfaces can be observed [46, 59, 208, 210, 213, 434, 520–528, 530].

The characteristic scale size and morphology of the structures formed depend on laser wavelength, insofar as it influences the ablation threshold and the number of irradiating pulses. These microstructures can be divided according to their mode of formation.

(i)   *Incident intensity* – conical structures appear on ablated polyimide film (amorphous phase) (Fig. 8.10) and poly(ethylene-2,6-naphthalate) (Fig. 8.11) due to shielding of the underlaying polymer by particulate debris (Fig. 8.12) formed by exposure at low laser fluences near the ablation threshold (E_t). These cones disappear on laser irradiation at intensities high enough to etch polyamide film (fluence $\gg E_t$) [208, 210, 213].

(ii) *Penetration of irradiation* – poly(methyl methacrylate) film doped with a high concentration of pyrene shows well-defined ablation. Small curd-like particles are observed for a film with a low dopant concentration. The origin of these particles can be explained by the low density and inhomogenous distribution of bond scission in the polymer subsurface during ablative photodecomposition with an increase in beam penetration [483].

(iii) *Phase in polymer* – semicrystalline polymer films of poly(ethylene terephthalate), poly(ethylene-2,6-naphthalate) and poly(phenylenesulfide), which are stressed biaxially or monoaxially, always show a wavy surface structure (Fig. 8.13) of the order of 1–50 μm after ablative photodecomposition, whereas amorphous films of these polymers produce a smooth surface. Mechanisms for the formation of these structures have been proposed based on selective etching of the amorphous and crystalline regions in the polymer and on a convection

Laser radiation

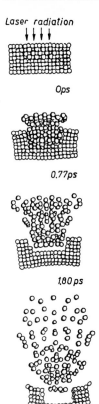

Ops

0.77ps

1.80ps

3.59ps

Fig. 8.8. Photochemical model of the movement of monomers as a function of time [269]

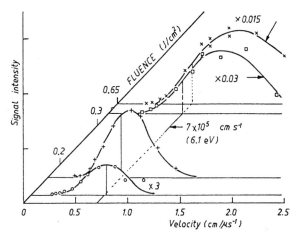

Fig. 8.9. Plot of the amplitude and velocity of the bandhead signal in the ablation of poly(methyl methacrylate) with 248 nm pulses [712]

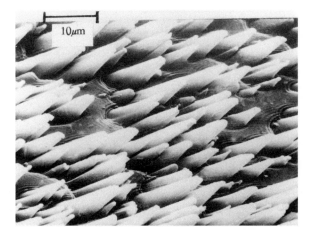

Fig. 8.10. Cones formed in polyimide film irradiated using the ArF laser [208]

instability in the heated surface layer. It is likely that stress relaxation in a melted surface layer is the dominant effect [46, 59, 434, 530]. In the case of poly(ethylene terephthalate) this wavy surface structure gives highly granular texturing (Fig. 8.14).

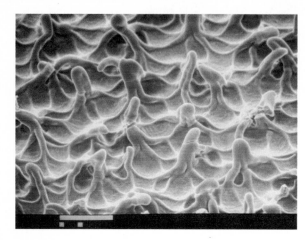

Fig. 8.11. Cones formed in biaxially stretched poly (ethylene-2,6-naphthalate) film exposed to 351 nm laser radiation [523] (provided by Dr H. Niino, National Chemical Laboratory for Industry, Tsukuba, Japan)

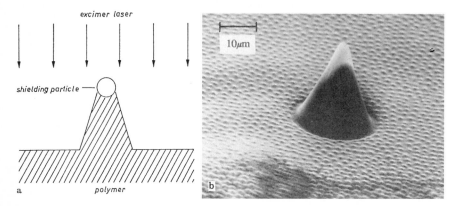

Fig. 8.12. a Schematic representation of the cone formation by ablation-resistant particle which shields the underlying polymer. **b** Cone formed in polyimide (unseeded) with normal-incidence irradiance [208]

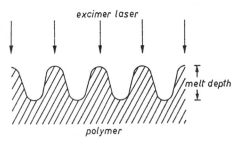

Fig. 8.13. Schematic representation of the rectangular pattern formation [208]

Fig. 8.14. Rectangular pattern formed in the biaxially oriented poly(ethylene terephthalate) film [208]

(iv) *Stress field* – mechanical high stress levels (approximately 100 MPa) lead to a textured morphology on the ablated polyimide surface due to microcrack formation and linkage of this damage during irradiation [755].

(v) *Surface scattering* – a ripple pattern appears on amorphous poly(ether sulphone) and polyimide films after a single exposure with a plane polarized beam with fluence in the range E_t to approximately $3E_t$ [209]. This is explained by surface scattering models (Fig. 8.15) which relate the surface roughness and the coherence of the incident beam [697]. This pattern is restricted to grating-like structures (Fig. 8.16), and its spacing is dependent on the wavelength of the laser line and incident angle to the film.

The degree of polymer crystallinity may be controlled by using control laser-induced texturing. Laser etching improves adhesion to polymers because texturing increases surface area.

Oriented, semicrystalline poly(ethylene terephthalate) tends to texture after multiple laser pulses more than amorphous, unoriented parts of a polymer matrix [206, 530]. Texturing is a result of the different etching rates between the amorphous and cyrstalline areas. Laser etching of semicrystalline polymer fibres, however, suggests that the formation of texture is the result of polymer relaxation [58].

UV laser radiation treatment of polymer surfaces (e.g. poly(ethylene terephthalate), polyethersulphone or polyimide) increases hydrophilicity on exposed surfaces during imagewise wetting and metallization [340, 341].

UV laser ablative photodegradation has found numerous micro-lithographic [86, 98, 217, 650], microelectronic [55, 56, 709], medical and surgical [454, 711, 796] applications.

A variety of structures can be produced in thin-deposited, free standing or bulk polymer films by excimer laser ablation. The pattern can be defined by a mask containing the designed pattern (Fig. 8.17).

The most technological application of laser ablation is patterning of polyimide film [100, 816], which is important industrially because of its toughness and electrical properties. The high resistance of this polymer towards chemicals and heat makes it difficult to pattern by any conventional method.

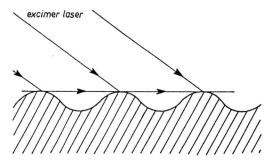

Fig. 8.15. Schematic representation of the grating-like pattern formation based on interference between the incident and surface-scattered waves. If interference produces a local fluence minimum at the peaks then selective etching enhances the pattern [208]

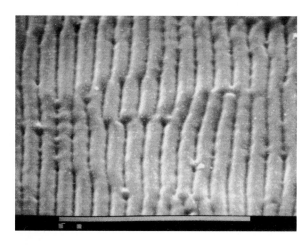

◄

Fig. 8.16. Grating-like pattern formed on poly(ether sulphone) surface exposed to 308 nm laser radiation [522] (provided by Dr H. Niino, National Chemical Laboratory for Industry, Tsukuba, Japan)

Fig. 8.17. Shallow etch crater produced in poly(etherether ketone) by XeCl laser ablation. The line is due to a fine wire placed on the circular mask to assist focusing [208] (provided by Dr G.A. Oldershaw, University of Hull, UK)

▼

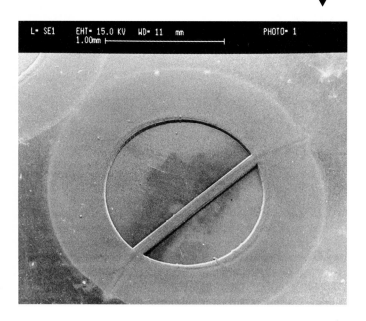

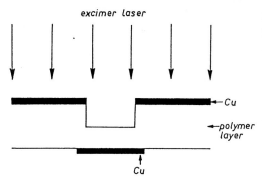

Fig. 8.18. Schematic diagram of excimer laser drilling of interconnect holes in multilayer printed circuit board [56, 208]

One of the most important applications is the laser drilling of vias (holes) (Fig. 8.18) in polyimide layers used in semiconductor device packaging [56, 208]. These holes are needed to allow connections to be made to underlying circuitry. Holes opened in the copper layers using conventional lithography define a mask for the excimer ablation (Fig. 8.19). Up to 30 000, 80-μm diameter, 65-μm deep vias between adjacent layers are produced, the laser addressing 60 of these at any one time.

An important advantage of UV laser ablation (in contrast to an infrared laser) is in the lack of carbonization at the new surfaces that are formed, because a surface film of carbon alters the electrical properties of the film in an undesirable way.

Fig. 8.19. Hole drilled in boundary of plastic contact lens using an ArF laser [208]

8.2 Photokinetic Etching

Photokinetic etching [713–716] uses a continuous beam from an argon-ion laser which is chopped with a rotating wheel, giving pulses of 50–500 μs. The polymer surface is moved with respect to the beam. The observed etching process is similar to that obtained in ablative photodecomposition, where pulses are 1 μs. However, longer 50 to 500-μs UV pulses cause a thermal process of etching by the internal conversion of the photon energy and the heating of the polymer to its pyrolysis temperature. Using this method a resolution of 10–100 μm can be obtained. A practical way to use the etching of polymers by microsecond UV pulses (e.g. from the argon-ion laser) has been found in cutting and etching processes.

8.3 Thermal Processing of Polymers by Visible and Infrared Laser Radiation

Until UV laser radiation, visible [518] and infrared [144] laser radiation can only be used for thermal processing of polymers, e.g. drilling (Fig. 8.20), cutting (Fig. 8.21), grooving (etching) (Fig. 8.22) and for welding [579, 742, 743].

During infrared CO_2-laser irradiation of different polymers, essentially all the absorbed energy is transformed into thermal energy, which causes melting, vaporization and flash pyrolysis of a polymer [99, 150, 211, 545].

Thermal degradation occurs from an accumulation of sufficient vibronic energy whereby main-chain bonds are broken with the eventual formation of molecular fragments sufficiently small to escape as volatile products. In Table 8.1 the energies required to ablate 1.0 g of a given polymer using CO_2 10.6-μm laser radiation are shown.

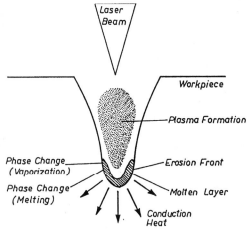

Fig. 8.20. Schematic representation of a laser drilling [144]

Polymer	Observed ablation energy, (kJ g^{-1})
Poly(α-methylstyrene)	2.4
Poly(tetrafluoroethylene)	3.8
Polystyrene	3.9
Poly(methyl methacrylate)	3.5
Poly(ethyl methacrylate)	3.4
Poly(n-butyl methacrylate-co-methyl methacrylate)	3.6
Poly(butadiene)	3.4
Epoxy	4.3
Polycarbonate	13.3
Polyimide	13.3
Poly(phthalocyanine) C–10	37

Table 8.1. Ablation energies of polymers (10.6 μm laser radiation at 55 W cm^{-2}) [150]

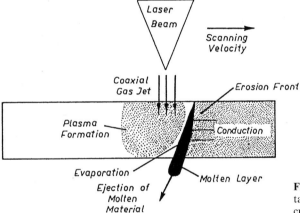

Fig. 8.21. Schematic representation of a laser through-cutting [144]

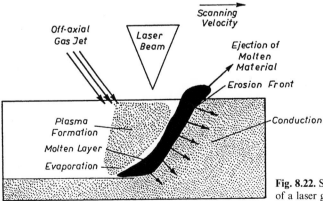

Fig. 8.22. Schematic representation of a laser grooving [144]

The short ablation may change only surface properties and morphology [538]. Ruby laser radiation causes destruction of polymers with crystalline structures [410, 415].

Several investigations have been performed on laser cutting and drilling operations for fibre-reinforced polymers, e.g. glass/polyester, graphite/polyester and aramid (Kevlar)/polyester materials [778–781].

Thermal processing of tissue by continuous wave and pulsed laser radiation (using argon, Nd:YAg and CO_2 lasers) is widely used in medicine, e.g. in the treatment of cancer, and tissue cutting control [796].

9 Practical Aspects of Polymer Photodegradation

9.1 Photodegradation of Polymers Under Weathering Conditions

Under natural (weathering) conditions several factors such as exposure to sunlight, day and night temperatures, seasonal variations, humidity and atmospheric contamination with highly corrosive agents in particular, are all important in the photodegradation of polymers and plastics. Other factors which may contribute to mechanical corrosion, e.g. gusts of wind loaded with dust and sand, impact of rain drops or of hail, freezing in winter, etc. may also be involved [176, 406, 609, 632, 637, 659, 809].

For many polymers the first step in the process that leads to a weathering-related failure is the formation of a brittle surface layer. Photo-oxidative degradation dominates at the surface because the UV intensity (from the sun's radiation) is highest there and because the approach of oxygen in the interior may be limited by diffusion whereas there is no shortage at the surface layer. Failure occurs by the propagation into the substrate of a crack formed within the brittle layer [585, 586, 637, 664, 665]. Further contributions to failure may be the development of tensile residual stresses which may lead to formation of microcracks in the embrittled surface layer without the need for any external load.

Initiation of failure without an external load can happen, additionally or alternatively, through the formation of flaws by the impact of wind-borne particles (e.g. sand) or rain. The localized degradation may occur preferentially in the vicinity of flaws that cause stress concentration within the residual stress field, thus enhancing the chance that they will nucleate a crack when stress levels are increased by applying an external force. Even a tensile stress of around $3\,MN\,m^{-2}$ would be expected to enhance the rate of photo-oxidative degradation in combination with a stress-intensifying flaw [586, 686].

The main weather parameters and their effects on the polymer, and the possibilities for and limitations on acceleration, are shown in Table 9.1.

9.2 Photodegradation of Polymers in Natural Waters

Photocatalytic oxidation may contribute to the formation of acid rain. Rainwater contains nitric and sulphuric acid, which may be formed by oxidation of nitrous and sulphurous acids. Nitrogen and sulphur oxides, NO and

Table 9.1. Possibilities and limitations for accelerated weathering of polymers [406]

Influencing Factors	Degradation Characteristics	Fundamental Possibilities of Acceleration	Limitations of Acceleration	Important Engineering Conditions for a Good Correlation
1. Sun radiation				
1.1. Spectral energy distribution	Type of degradation $= f(\lambda)$ (Quantum energy)	Altered spectral energy distribution in UV	Reactions not occurring outdoors, therefore	Radiation source Filter system reverse order possible
1.2 Irradiance	Rate of degradation $= f(I)$	More duration per day Higher irradiance	Degradation linear function of $I \times t$?	Adjustment and control of irradiance
2. Heat	Speed of degradation $= f(T)$ Type of degradation $= f(T)$ (Activating energy)	Higher temperature of samples	Different degradation mechanisms at different temperatures (thermal-ageing)	Adjustment and control of black panel temperature
3. Humidity Water	Swelling/Shrinking Diffusion-controlled process Participation in degradation reactions	More frequent dry-wet cycles	Duration of dry-wet periods $=$ hours	Adjustment and control of relative humidity More than 100% relative humidity not possible
4. Oxygen	Photooxidation Diffusion-controlled process	Higher partial pressure of oxygen		
5. Industrial Pollution	Direct reactions with the polymers with and without radiation Reactions with products of a photo-chemical reaction in the air	Higher concentration		Adjustment and control of different concentrations of pollution in the exposure area
6. Mechanical Stress	Faster degradation when influenced by # 1, 4 and 5	More mechanical stress		Sample under defined mechanical stress

NO_2 and SO_2 and SO_3, dissolve in water to form mixtures of nitric, nitrous, and sulphurous acids.

In natural waters, the predominant light-absorbing species are usually the humic substances, which are involved in the formation of [827]:

(i)   hydrogen peroxide by the following mechanism:

$$HA_{aq} \xrightarrow{h\nu} {}^1HA_{aq}(S_1) \longrightarrow {}^3HA_{aq}(T_1) \tag{9.1}$$

$${}^3HA_{aq} \longrightarrow HA_{aq}^+ + e_{aq}^- \tag{9.2}$$

$$e_{aq}^- + O_2 \longrightarrow O_2^-. \tag{9.3}$$

$$2O_2^-. + 2H^+ \longrightarrow H_2O_2 + O_2 \tag{9.4}$$

where HA_{aq} is the water solvated humic acid, and e_{aq}^- is the electron solvated;

(ii) singlet oxygen ($^1O_2$) [828]:

$${}^3HA_{aq} + O_2 \longrightarrow HS + {}^1O_2. \tag{9.5}$$

The superoxide anion (O_2^-) undergoes various kinds of reaction, for example oxidation, reduction, nucleophilic substitution and addition reactions with substrates [486]. A little attention has also been given to the role of superoxide anion in the degradation of polymeric materials [420, 550, 552–554].

In natural waters, nitrate and nitrite photolysis is an additional source in the production of hydroxyl radicals (HO·), and thus is a contributor to the natural abiogenic degradation of organic compounds in water [325, 504].

One important way for the oxidation of polymers in natural water could be UV (light) induced reactions with the hydrous amorphous Fe(III) phase [609].

9.3 Photodegradation of Polymers Under Marine Exposure Conditions

All polymeric and plastic materials exposed to the sea invariably undergo fouling. The initial stages of fouling result in the formation of a biofilm on the surface of plastic. Gradual enrichment of the biofilm leads to a rich algal growth within it. Consequently, the biofilm becomes opaque and the sunlight available to the plastic for photodegradation is restricted. Thus the rate of photo-oxidative degradation at sea might be determined in part by the rate of fouling [569].

Advanced stages of fouling are characterized by the colonization of the plastic surface by macrofoulants such as bryozoans. The weight of the macrofoulant and that of debris they entrap might even partially submerge the material. As the ultraviolet portion of sunlight is attenuated on passage

Table 9.2. Comparison of weathering data for exposure on land and at sea [569]

Sample	Duration of exposure	Per cent decrease in the mean value of tensile property			
		Air		Sea Water	
		Strength[a]	Extension	Strength[a]	Extension
Polyethylene film	6 months	19.1	95.0	5.2	2.0
Polypropylene tape	12 months	85.0	90.2	11.0	25.6
Netting	12 months	no change	no change	no change	no change
Latex balloons	6 months	98.6	93.6	83.5	38.0

[a] The percentages reported are based on the maximum load in the case of netting and polypropylene tape materials

through sea water, submerged plastics would necessarily undergo a slower rate of photodegradation.

Microbe-rich foulant film may, however, tend to accelerate the biodegradation process.

In Table 9.2 general findings are shown of the exposure study by comparison of tensile properties, before and after exposure in air and in sea water, of different polymers and plastics. Except for netting, rates of degradation for the samples in sea water are much slower than the degradation rates on land. Netting material did not show any significant variation in tensile properties due to the type or duration of exposure [569].

The marked retardation of the weathering process observed in some types of plastic materials floating in sea water might be attributed to the following.

(i) Differences in heat build-up. A significant fraction of the sunlight impinging on a plastic surface is absorbed by the material as heat. Depending on the nature of the plastic, the velocity of the air around it, and the temperature difference between the plastic and the surroundings, this absorbed energy maintains the plastic at a temperature as much as 30°C higher than that of the surrounding air [45, 268, 485]. At sea, however no such heat build-up occurs.

(ii) Fouling of samples in sea water. Samples floating on the sea undergo extensive fouling during the exposure.

9.4   Hydrolysis Processes During Ageing of Polymers

Hydrolysis of some polymers (e.g. polyesters) plays an important role in environmental ageing (degradation) of polymeric materials.

Hydrolysis is the most important process in low temperature degradation at temperatures above the polymer glass transition temperature (T_g). Under these conditions, moisture may more easily permeate the amorphous regions of the polymer and induce hydrolytic breakdown. Random hydrolysis is the dominant mechanism under high humidity conditions as evidenced by the comparable rates of formation of hydroxyl and carbonyl end groups. During the early stages of the low temperature hydrolytic ageing of a polymer there is an initial rapid hydrolytic annealing process due to the plasticization effect of the water, followed by a slower crystallization change due to chain scission and hydrolysis [16].

9.5 Plastic Ecological Problems

Millions of tons of litter and waste plastics are floating on rivers, lakes, and ocean thousands of miles from any city. Deposited in recreation areas and parks, on mountains and remote beaches, this has become a modern ecological problem. The collecting of used plastics in many undeveloped countries round the world is a great technical and economical problem [156, 440].

Plastic litter in the ocean environment increases from year to year. Commercial fishing activity is the primary source, and gear-related debris is often an important component of beach debris in regions near commercial fisheries. A small fraction of this gear (usually plastic netting) is inevitably lost (or deliberately discarded) at sea each year and ends up as persistent marine debris. Significant amounts of plastic packaging materials are also discarded into the oceans from other vessels. Plastic debris from land-based sources such as beach litter may also cause sea pollution. Persistence of plastic debris at sea poses several hazards to marine life, e.g. the entanglement of marine animals by net fragments, lengths of monofilament line, rope, packing bands, and six-pack carriers [60, 427, 583].

The synthetic polymers that compose plastics stubbornly resist dissolution by water (rain), air and sunlight and are unpalatable to the microorganisms that break down natural materials in the eco-system. Burning these plastic litters does not solve things either, as many common plastics (e.g. poly (vinyl chloride)) give off toxic gases when torched. Halogen-base fire retardants customarily added to many plastics quench flames but permit the plastic to smolder and smoke. Only a few plastics, and still on a very low-scale, are reusable or recovered [676, 732, 775, 776].

The solution of plastic ecological problems lies in the developing of photo- and/or bio-degradable polymers with controlled lifetimes [207, 378, 676, 681]. It is becoming increasingly apparent that the future of plastics (especially in the packaging industries) must be limited by the problem of accumulation of waste. The plastics industry is entering a new phase in taking responsibility not only in the production of high-tech plastics but also for their collection, disposal or reprocessing.

Recycling of polymer debris (Fig. 9.1) presented as clean and segregated waste seems to be a future solution of plastic ecological problems [676].

Examples of useful products that can be made economically are polyester insulation fibres from recovered poly(ethylene terephthalate) (PET) bottles, waste bags from car batteries and discarded crates etc. (Fig. 9.2) [676].

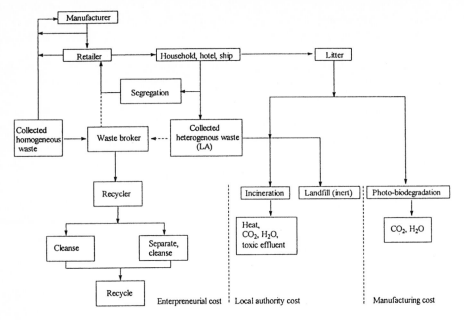

Fig. 9.1. Sources and fate of plastics waste and litter [676]

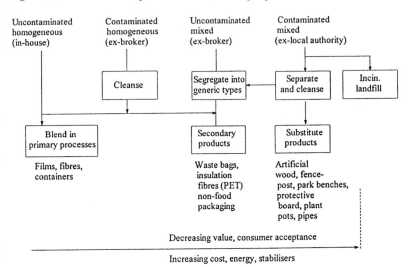

Fig. 9.2. Materials recycling [676]

9.6 Photodegradable Polymers with Controlled Lifetimes

Photodegradable polymers containing light absorbing groups (chromophores) in a chain and/or side groups or as additives (photoinitiators) have importance in the packaging industry and in agriculture application (mulching films).

Photodegradable polymers and their practical applications have been reviewed in numerous publications [292, 293, 310, 311, 316, 318, 378, 403, 436, 547, 565, 666, 667, 669, 672, 674, 676, 677, 684].

Polymeric packaging materials require not only a polymer that will degrade, but one that will degrade at a controlled and predictable rate, and will produce ecologically acceptable products. Introducing light absorbing groups into a polymer structure and/or as additives into plastics causes absorbed sunlight radiation to initiate photo-oxidative processes, the chain to break and formation of smaller segments. Since the physical and mechanical properties of a polymer (plastic) depend on the length of the chain (molecular weight and molecular weight distribution), if the chain is broken it will become very fragile and easily be mechanically disintegrated into small fragments. Microorganisms tend to attack the ends of macromolecules and the number of ends is inversely proportional to the molecular weight. Generally, in order to make polymers (plastics) degradable, it is necessary first to break them down into very small particles with large surface area, and second to reduce their molecular weight. Degradation in a landfill environment is primarily anaerobic and does not occur evenly throughout the landfill. While bio-degradation does occur, it occurs very slowly. For example, even food wastes have been found in recognizable form after many years in a landfill.

A serious obstacle in the developing of polymers with controlled lifetimes is the matter of how to achieve an accurately definable degree of decomposition for different climatic and other conditions, when the rate of degradation of degradable plastics will vary in different regions as a result of the varying sunlight exposure, day/night temperature, humidity and activity of locally present bacteria and other microorganisms.

Another important problem is the fact that the use of photodegradable plastics would raise the cost of packaging and hence reduce their competitiveness.

While environmentalists applaud introduction of photo- and/or bio-degradable polymers some industries and customers may baulk. The whole notion of degradability, in fact, seems to contradict what the plastic industry has preached for years: that plastics are not fragile throwaway goods. The object of making plastics has been to make them durable.

There are a number of other concerns about the use of photo-degradable polymers. One is that the inadvertent incorporation of photodegradable plastics into the stream of recycled plastics could lead to accelerated degradation of the resultant products.

Still very little is known about how photodegradable polymers perform in different environmental settings [156].

(i)   Will the use of photodegradable polymers pose an ingestion threat to wildlife?
(ii) Will the degradation byproducts (i.e. the small pieces of undegraded plastic) pose a threat to a wildlife (e.g. their entanglement or strangulation) in the residue of photodegradable films?
(iii) Will the use of photodegradable polymers increase littering?

A controversial question is whether or not degradable polymers can disappear without leaving intermediates and residues which may be a source of air, water and soil pollution. Disintegrated parts of plastic packages can be easily transported by wind and water kilometers away from the point of their degradation.

There are three general areas of concern regarding the potential environmental hazard of photodegradable plastics.

(i)   Is the polymer itself more toxic due to its enhanced photodegradability (e.g. by using photoinitiators)?
(ii) Are the byproducts of the photodegradation process toxic, carcinogenic and mutagenic?
(iii) Does the degradation process increase the leachability of additives from the polymers?

Industry producing photodegradable polymers must generate and make available data such as:

(i)   rates of degradation in different environments;
(ii) identification of chemical and physical degradation products and their impacts (including air emission);
(iii) possible impacts on recycling, assuming that the use of this management technique expands;
(iv) leachability of additives from products during the degradation;
(v)   toxicological tests.

9.7   Ecolyte Plastics

Ecolyte plastics are based on polymers (copolymers) containing carbonyl groups, which absorb the ultraviolet radiation of solar emission, but do not absorb visible light [310, 487, 641].

Consequently they do not contribute to the colour, nor does the degradation process occur by the action of visible light. Depending on the structure of the carbonyl group, and its localization along the polymer chain, the efficiency of its reaction to ultraviolet radiation can be changed. The rates as measured by quantum yields are independent of oxygen, moisture, temperature and pressure, and consequently the rate can be predicted from the intensity of the ultraviolet radiation and the time of exposure only. The rate of degradation is directly proportional to the concentration of carbonyl groups for a given thickness of a plastic specimen. In order to provide an acceptable rate of degradation for polystyrene (Ecolyte S), it is necessary to include less than 1% of carbonyl groups in the polystyrene molecule.

The chain process begins as soon as the plastic is exposed to solar radiation, although there is a certain period of time necessary before an appropriate change in the physical properties occurs (Fig. 9.3). The reason for this is that, above a certain molecular weight, which is sometimes called the "critical molecular weight", there is only a small change in the physical properties of

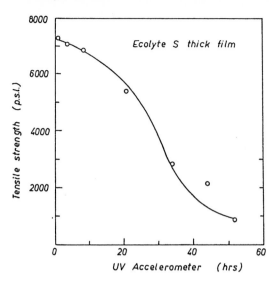

Fig. 9.3. Decreasing of tensile strength of Ecolite S (copolymer styrenecarbon monoxide) during irradiation in UV accelerometer [310]

the polymer as the molecular weight changes. Once this molecular weight is reached, however, any subsequent decrease in molecular weight will cause a drastic change in the properties of the material. This means that, even after exposure to solar radiation, the plastic material will still retain its useful properties for a certain period of time, and this time can be controlled at will in the manufacturing process.

The photodegradation process works by the Norrish Type I and Type II processes (cf. Sect. 4.23) with a wide variety of vinyl polymers, including polyethylene (Ecolyte E), polypropylene (Ecolyte P), polystyrene (Ecolyte S), poly (methyl methacrylate), polyacrylonitrile, poly(methacrylonitrile) and their copolymers. In each case, a substantial rate of degradation is achieved with minor amounts of ketone groups.

The ketone groups is normally introduced into the polymer during the manufacturing process, usually by copolymerization with another monomer containing the group in the appropriate position. With vinyl monomers this is usually done by copolymerization with a vinyl ketone monomer. In condensation polymers such as polyamide and polyesters, the group is introduced by synthesizing a difunctional monomer containing the ketone group and adding it during the preparation of the polymer. In some cases, ketone groups can be introduced by a chemical post-treatment. These compositions and processes have been patented by the University of Toronto. The commercial development of these has been undertaken as a joint enterprise (or under licence) by several companies: Eco-Plastics Ltd, Toronto (Canada); Dow Chemical, Midland (USA); Dupont Co. Willmington (USA); Union Carbide, Danbury (USA) and Royal Packaging Industries Van Leer, Amsterdam (Holland).

9.8   Plastic Protective Films

Plastic protective films (mulching films), which contain photoinitiators to accelerate their photodegradation, are widely used in the "sunbelt" countries, which include the whole of southern Europe, Africa, the Middle- and Far-East, USA, Mexico and China, to control crop growth and increase yields [291–293].

The only commercial time-controlled photo-degradable mulching film currently in use is based on such a photoinitiator, iron(III)dialk-yldithiocarbamate, which at low concentration causes catastrophic photo–oxidative degradation of a polymer, according to the following reactions [38–42, 44, 293, 497, 668, 670, 676]:

$$\left[R_2NC \underset{S}{\overset{S}{\diagup}}\diagdown \right]_3 Fe$$

$$R_2N - C \underset{S\cdot}{\overset{S}{\diagup}} \xrightarrow{+PH} R_2N - C \underset{SH}{\overset{S}{\diagup}} + P\cdot \qquad (9.6)$$

$$+POOH$$

$$R_2NC = S + SO_3 + Fe^{3+} \qquad (9.7)$$

$$+POOH$$

$$Fe^{3+} + PO\cdot + OH^- \qquad (9.8)$$
$$Fe^{2+} + POO\cdot + H^+ \qquad (9.9)$$

This photoinitiator introduced into polyethylene can lead to very short-lived materials such as packaging which degrade in the environment almost immediately they are discarded, protective films for annual crops where a lifetime of several months is required, protective mulch for young trees which may be required to stand for 3–4 years in exposed environments before embrittlement, etc.

The commercial development of these mulching films has been undertaken by EniChem Agricultura (Italy); Plastopil (Israel); Polydress (Germany); Plastigone Technologies Inc. (USA); and Amerplast (Finland).

The serious disadvantages of photodegradable mulching films are

(i)   increased litter and ecological problems – the considerable amount of non-decomposed film can remain in soil, and incompletely photo-degraded pieces of film are blown by the wind over large areas (specially accumulated on bushes and trees);
(ii) increased accumulation of metals in the soil, which can have as yet unknown effects on the genetic functionality of plants and crops;
(iii) increased accumulation of metallo-organic photoinitiators (extracted by the rain) and photolysis products in the soil, which can have toxic, mutagenic and allergic effects.

Microorganisms present in soil tend to attack the remaining parts of photo-disintegrated particles of a film and finally produce H_2O and CO_2.

9.9   Recovery of Photodegraded Polymers

Physical degradation (morphological changes, cracking, breaking of physical bonds) can be reversed by reprocessing the polymer. On the other hand, chemical changes (chain scission, crosslinking, formation of oxidized groups) are virtually irreversible. However, both physical and chemical degradations are highly heterogenous processes and reprocessing can disintegrate the do-mains of locally degraded material.

Because photo-oxidative degradation dominates at the surface layer, it has been suggested that restoration of a polymer sample can be achieved, first by removing (e.g. by sanding) the embrittled top layer, and then by protecting the scoured surface by a very resistant UV-cured coating. The overall recycling process of photodegraded poly(vinyl chloride) is presented in Fig. 9.4 [184].

9.10   Photodegradation of Polymers
in the Polluted Atmosphere

The atmosphere above sea level consists of troposphere (first 10 km) and stratosphere (10–50 km). The troposphere contains 75% of atmosphere mass, and together with the stratosphere more than 99%. The atomosphere at sea level consists of nitrogen (75.527%), oxygen (24.143%), argon (1.282%), car-bon dioxide (4.5–6×10^{-2}%), neon (1.25×10^{-3}%), krypton (3.3×10^{-4}%), he-lium   (7.25×10^{-5}%)   and   hydrogen   (3.48×10^{-6}%)   [833].   Space   gas concentration (number density) decreases with increasing altitude above sea level (Fig. 9.5).

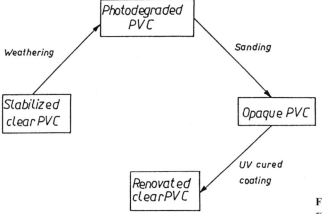

Fig. 9.4. Performance of renovated polymers [184]

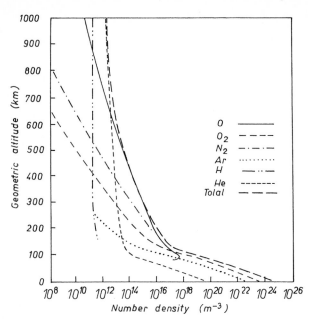

Fig. 9.5 Space gas concentration [730]

The main components of atmospheric environmental pollutants in the troposphere are both sulphur oxides – sulphur dioxide (SO_2) and sulphur trioxide (SO_3), sulphuric acid (H_2SO_4), sulphates (SO_4) and nitrogen oxides – nitric oxide (NO) and nitrogen dioxide (NO_2) [235, 301, 441, 640].

Chemical reactions play a very important role in the manner in which cities and industries contribute to the particle loading of our atmosphere. To a certain extent, chemical reactions are important in smog of the coal-burning type, since sulphur dioxide is continually being oxidized up to sulphur trioxide which is then hydrated to form sulphuric acid droplets.

Sulphur dioxide photosensitizes formation of singlet oxygen ($^1O_2$) in the atmosphere [173, 174]:

$$SO_2 \xrightarrow{h\nu} SO_2^* \tag{9.10}$$

$$SO_2^* + O_2 \longrightarrow SO_2 + {}^1O_2(^1\Sigma_g^+) \tag{9.11}$$

Reaction of sulphur oxides with polymers has been discussed in detail elsewhere [257, 336]. The reaction of SO_2 with photo-oxidized polymers decreases the extent of observed oxidation, because SO_2 participates in the decomposition of polymeric hydroperoxide groups [19, 20]:

$$POOH \xrightarrow{SO_2} POSO_2^* + HOSO_2{\cdot} + RSO_2{\cdot} \tag{9.12}$$

$$POO{\cdot} \xrightarrow{SO_2} POOSO_2^* \longrightarrow PO{\cdot} + SO_3 \ . \tag{9.13}$$

However, acceleration of photo-oxidation of different polymers due to SO_2 has also been reported [362, 367, 370, 372, 375].

It has been established in a series of experiments that the combined action of SO_2, water and UV radiation is responsible for rapid fading. On exposure to radiation solution of sulphuric acid is formed on the surface of the specimen from an aqueous SO_2 solution and cause fading, principally due to reaction with pigment.

The most important triggering reaction in photochemical smog (resulting from car pollution in many heavy traffic cities like Los Angles, Mexico City, Cairo and many others) is photochemical decomposition of nitrogen dioxide (NO_2) to form nitric oxide (NO) and atomic oxygen (O) [205, 235, 441]:

$$NO_2 \xrightarrow{h\nu} NO + O \ . \tag{9.14}$$

Atomic oxygen can react with molecular oxygen to form ozone, but ozone also reacts with nitric oxide to form nitrogen dioxide and oxygen [235, 301, 441, 640]:

$$O + O_2 \longrightarrow O_3 \tag{9.15}$$

$$NO + O_3 \longrightarrow NO_2^* + O_2 \tag{9.16}$$

$$NO_2^* + O_3 \longrightarrow NO_3 + O_2 \tag{9.17}$$

$$NO_2^* + O \longrightarrow NO + {}^1O_2 \tag{9.18}$$

$$NO_2^* + O_2 \longrightarrow NO_2 + {}^1O_2 \tag{9.19}$$

where NO_2^* is the electronically-excited state.

Because of the very complicated mechanisms of photoreactions of nitroxides in the atmosphere, it is difficult to select their specific reactivities with polymers. However, some reactions of NO_2 with polymers have been reported elsewhere [365–368, 370, 375, 379].

9.11 Photodegradation of Polymers in the Stratosphere

The stratophere is also filled with trace of many reactive gases (Fig. 9.6). Gases emitted in the troposphere that migrate to the stratosphere are N_2O and, to a lesser extent, NO and NO_2, water and methane, chloroform, bromoform and chlorofluorocarbons. These stable gases are broken down in the stratosphere by either sunlight or chemical products of sunlight and become the reactive species that interact with each other but also destroy ozone [106]. Figure 9.7 shows the long-lived source gases (as solid bars), and the range of the resulting reactive gases, including everything from free radicals to acids (given by the shaded bars).

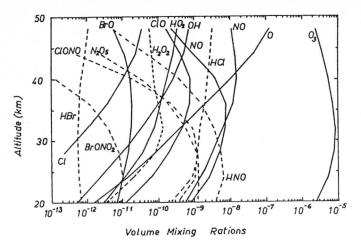

Fig. 9.6. Calculated altitude distributions of the volume mixing ratios of several trace gases. The relative trace gases that directly affect ozone are given by *dark lines* [106]

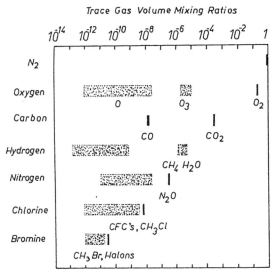

Fig. 9.7. Abundances of gases in the stratosphere [106]

The production of ozone occurs by the reaction

$$O_2 \xrightarrow{h\nu(UV)} 2\,O \tag{9.20}$$

$$O_2 + O \longrightarrow O_3 \tag{9.21}$$

whereas the destruction of ozone occurs by a number of mechanisms that results in:

$$O_3 \xrightarrow{h\nu(\lambda 320\,nm)} O + O_2 \tag{9.22}$$

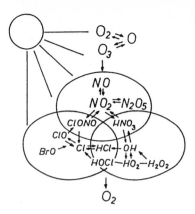

Fig. 9.8. Schematic of the major gas-phase cycles in the stratosphere [106]

$$O + O_3 \longrightarrow 2\,O_2 \tag{9.23}$$

$$O + H_2O \longrightarrow 2\,HO\cdot\;. \tag{9.24}$$

In the mid-latitudes (20–50 km above sea level) ozone photodecomposition is accelerated by nitrogen oxides and inorganic halogen chemicals (Fig. 9.8).

Hydroxy radicals (HO·) are apparently the atmosphere's most reactive species [537]. A single HO· radical in a clean atmosphere survives 1 s but only 1 ms in polluted regions. These life-times are inversely proportional to the concentration and HO· reactivities of those species with which HO· reacts. Atmospheric CO is the dominant reactant with HO· radicals and methane is second, whereas in polluted air an immense array of anthropogenic and biogenic hydrocarbons dominate HO· radical removal. The chemical lifetime of an individual HO˙ radical is given by

$$(\Sigma k_i T_i)^{-1} \tag{9.25}$$

where T_i indicates the concentration of an atmospheric trace gas that reacts with HO˙; and k_i is the rate constant.

Photodegradation of polymers in the stratosphere is important in the design aircraft, rockets, space vehicles, etc. However, results from research in this field are not available yet. Most of the knowledge on this problem has been collected from experiments done under artificial laboratory conditions, e.g. reaction of atomic oxygen with polymers.

9.12 Photodegradation of Polymers in Space

Most polymers suffer a loss of surface integrity when exposed to the space environment. This effect is generally attributed to chemical erosion by atomic oxygen, which is faciliated by the high contact energies ($\approx 5\,\text{eV}$) for atomic collisions with forward-facing spacecraft components. The practical im-

Table 9.3. Solar variability (NASA data) [730]

Spectral region	Wavelength	Flux $(J\ m^{-2}\ s^{-1}\ \mu m^{-1})$	Variability
Radio	$\lambda > 1$ mm	10^{-11}–10^{-17}	$\times$ 100
Far infra-red	1 mm $\geq \lambda > 10\mu m$	10^{-5}	Uncertain
Infra-red	10 $\mu m \geq \lambda > 0.75\mu m$	10^{-3}–10^{2}	Uncertain
Visible	0.75 $\mu m \geq \lambda > 0.3\mu m$	10^{3}	$< 1\%$
Ultraviolet	0.3 $\mu m \geq \lambda > 0.12\mu m$	10^{-1}–10^{2}	1–200%
Extreme ultraviolet	0.12 $\mu m \geq \lambda > 0.01\mu m$	10^{-1}	$\times$ 10
Soft X-ray	0.01 $\mu m \geq \lambda > 1$ Å	10^{-1}–10^{-7}	$\times$ 100
Hard X-ray	1 Å $\geq \lambda$	10^{-7}–10^{-8}	$\times$ 10–$\times$ 100

portance of the surface erosion of polymers subjected to atomic oxygen O(^{3}P) are important for design of space shuttles and space stations.

Spacecraft (shuttles, satellites, launch vehicles) operating in the space environment are exposed to a variety of hazards.

(i) Atomic oxygen O(^{3}P), which is produced from dissociation of molecular oxygen ($^3\Sigma_g^-$) by UV solar radiation at the average density of 5×10^8 atom cm^{-3}:

$$O_2(^3\Sigma_g^-) \xrightarrow{h\nu} 2\ O(^3P)\ . \tag{9.26}$$

Due to the high velocity (around 8 km s^{-1}) required to maintain a spacecraft in orbit, the collision energy for impacts between atomic oxygen and spacecraft surfaces is about 5 eV. The flux of oxygen atoms to the surface (3×10^{14} atom cm^{-2}s^{-1}) is a function of orbital altitude, orientation to the surface relative to the flux and phase of the solar cycle [574].

(ii) Ultraviolet and extreme ultraviolet, soft and hard X-rays (cf. Table 9.3) which cause photo-, ablative- and radiation degradation.

(iii) Charged particles, electrons and protons which have energies 0.4–4 MeV (and each has a flux of around 10^8 cm^{-2} s^{-1}), which can cause the high energetic radiation degradation [296, 399, 658]. Electrostatic charging of a spacecraft travelling through the near-Earth environment occurs when it enters or leaves the Van-Allen radiation belts (Fig. 9.9). As a consequence, currents will occur between the space vehicle and the plasma, imbalance of which will cause the spacecraft to develop a charge. The two major sources of currents are the ambient plasma itself and photoelectron emission due to sunlight, and in particular the short wavelength component of this radiation [730].

(iv) Meteoroids, micrometeoroids and space debris. These are solid objects whose mass and size vary over many orders of magnitude. Thin (or even thick) polymer films are usually penetrated by these impacts.

(v) Thermal shock resulting from the fact that the spacecraft in synchronous orbit is periodically eclipsed from the sun by the Earth. The dominant radiative heat input due to solar radiation in the near–Earth environment is 137±5 Wm^{-2}. Secondary input occurs due to Earth albedo (the re-

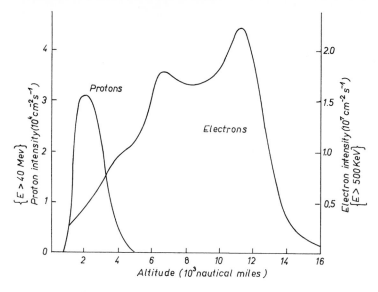

Fig. 9.9. Van Allen radiation belts (idealized) [730]

flection of solar radiation from the top of the atmosphere) and Earth shine (the black-body radiation of the Earth), and has a magnitude of around 20 W m^{-2} [730].

In Table 9.4 some information on the stress/performance characteristics of materials used for the construction of spacecraft vehicles is collected [494].

In order to study the effects of the space environment on different materials including polymers [147, 731], NASA's Office of Aeronautic and Space Technology launched in 1984 the Long Duration Exposure Facility (LDEF) unmanned, free-flying spacecraft (Fig. 9.10) in a geosynchronous Earth orbit (GEO) (low Earth orbit for long term (5 years and 9 months, approximately 34 000 orbits or three quarters of a billion mile journey) exposure to the space environment) [401, 548]. The LDEF spacecraft was a 12-sided, aluminium open grid frame, 4 m in diameter and 12 m long (weighing about 11 tons), to the surface of which were attached different materials submitted to the space environment.

Specimens selected for this study came from NASA Langley Research Center (Hampton, VA), Polymeric Materials Branch experiments and from materials made available to the Space Environmental Effects on Materials Special Investigation Group (MSIG) during LDEF desintegration activities at the Kennedy Space Center in the January–March 1990 time period.

The materials examined were silvered fluorinated ethylene-propylene teflon blanket material, polysulphone matrix resin/graphite reinforced composite, Kapton film (poly(N,N'-(p,p-oxydi-phenylene) pyromellitide, epoxy polymer matrix composite, and several high performance polymer films [104, 105, 143, 446, 490, 817, 818]. Long term exposed polymeric samples show loss of surface integrity and surface erosion. The UV (vacuum) and high energetic radiation

Table 9.4. Stress/performance characteristics of materials [494]

Parameter / Material	Main categories	Vacuum	Particulate radiation	Ultraviolet radiation	Temperature High	Low	Thermal cycling
Adhesives	Epoxies, phenolic, polymethanes, silicones, cyano-acrylates	Outgassing	Outgassing increased	Optical adhesives darken. Outgassing increased	Causes degradation 300°C max. for poly-imide epoxy 170°C max.	Hardening embrittlement	Unmatched expansion coefficients cause failures
Glasses	Silicates, sapphires, fluorides, poly-styrene, acrylics, etc.	Contamination danger only	Most harmful (10^3 Rad.)	Harmful	Thermal shock main problem		
Lubricants	Hydrocarbons, silicones esters, MoS_2, WSe_2, Ph	Evaporation, dry-off creep contamination	Metal screening usually effective		Accelerated evaporation etc.	No major problem	
Plastic films	Polyolefins, poly-esters, fluorinated plastics, etc.	Film stiffeners: contamination	Deformation embrittlement discolouration		Degradation	Embrittlement	Damaging to metalized films, metal film detachment
Potting compounds	Epoxies, silicones, polymethanes	Contamination corona	Minor problem		Chemical degradation	Shrink rigid increase; internal stresses increase.	Cracking debonding

Parameter / Material	Main categories	Vacuum	Particulate radiation	Ultraviolet radiation	Temperature High	Low	Thermal cycling
Reinforced and thermosetting resins	Epoxies, phenolics, melamine, polyesters	Outgassing corona contamination	No problem				can be a problem due to anisotrophy of reinforced plastics
Rubbers	Polybutadiene, poly-chloroprene, acrylics, nitrile etc.	Outgassing from additives and de-polyurization of base polymer. Contamination	Will harden or soften		Decomposition	Hardening, stiffening crazing, crushing	See high/low temperature
Thermo-plastics	Polyamides, acetal, polyolefins, acrylics, etc.	Degradation due to outgassing of stabilizing additives. Contamination	Discolouration, outgassing, hardening		Softens	Hardening, embrittlement	
Paints	Epoxy, silicone etc. binders, ZWO, TLO_2, Al, C for white/black pigment	High outgassing	Absorbance severely affected; embrittlement		Degradation, paint flaking	Minor problem	Degradation of paint flexibility

Fig. 9.10. NASA Long Duration Exposure facility, unmanned free-lying spacecraft (provided by NASA, USA)

combines with atomic oxygen $O(^3P)$ to initiate degradation (chain scission and crosslinking), which greatly affects the polymer's structural properties.

In Table 9.5. prohibited and non-preferred materials for the construction of spacecraft vehicles are listed [494].

9.13 Atomic Oxygen Reactions with Polymers

Atomic oxygen $O(^3P)$ reacts with almost all polymers (PH) causing surface erosion (etching and material mass loss) [234, 274, 300, 329–331, 344, 431, 461, 469, 606, 609, 649, 683, 798]:

$$PH + O(^3P) \longrightarrow P\cdot + \cdot OH \qquad (9.27)$$

$$P\cdot + O(^3P) \longrightarrow PO\cdot \qquad (9.28)$$

$$PO\cdot + O(^3P) \longrightarrow POO\cdot \ . \qquad (9.29)$$

Table 9.5. List of prohibited and non-prohibited materials [494]

1.	All adhesives must be 100% solid
2.	Polyvinychloride backing tapes
3.	Cellulose, paper, fabric, etc.
4.	Varnishes and coatings which rely on solvent evaporation for hardening
5.	Canada balsam; organic glasses in high-precision equipment
6.	Direct space exposure of most oils and greases
7.	Graphite–is an abrasive in vacuum
8.	Cadmium, zinc (whisker growth)
9.	Paints should be avoided where possible
10.	Polyvinychloride and acetate; cellulose and acetate, plastic films
11.	Potting should be avoided where possible
12.	Polyester laminates
13.	Polysulphide rubbers; rubbers containing plasticizers; chlorinated rubbers
14.	Polyvinyl chloride (PVC) thermoplastic, polyvinyl acetate butyrate many polyamides.

The etch rate (the reaction probability) (R) is expressed by the equation

$$R = AE^n \qquad (\text{amu atom}^{-1}) \tag{9.30}$$

where A and n are constants; E is the impact energy (eV), and $1\,\text{amu} = 1.66 \times 10^{-24}\,\text{g}$.

In Fig. 9.11 the log-log plot of etch rate vs $O(^3P)$ atom impact energies for Kapton polymer (poly(N,N'-(p,p'-oxidiphenylene) pyromellitide) is given. The

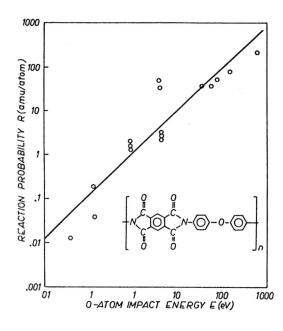

Fig. 9.11. Effect of atomic oxygen (○) impact energy on the etch rate of poly (N, N'-(p,p)-oxydiphenylene) pyromellitide (Kapton) [300]

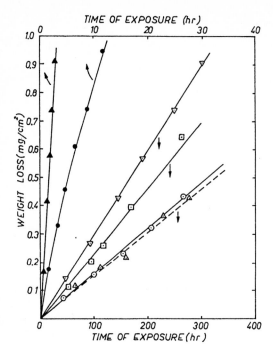

Fig. 9.12. Typical kinetic plots for atomic oxygen O(^{3}P-induced weight loss in various polymer films: (▲) poly (vinylidene fluoride); (●) high-density polyethylene; (▽) polystyrene; (□) poly (ethylene terephthalate); (○) poly(vinylidene fluoride): (△) poly(N, N'-$(p,p'$-oxydiphenylene) pyromellitide (Kapton) [300]

straight line plot in Fig. 9.11 can be described by Eq. (9.30), with A = 0.87 and n = 0.92 [51, 233, 300, 830].

The kinetic plots for weight loss or surface erosion in some of the polymer films exposed to O(^{3}P) downstream from the plasma glow in the laboratory reactor are shown in Fig. 9.12.

Mass loss rate was found to be directly related to the exposure area and to be independent of sample thickness. Comparative mass loss data is shown in Table 9.6.

Table 9.6. Material mass loss rate [798]

Polymer	Rate of mass loss 10^{-6} (kg h^{-1}/area)
Polytetrafluroethylene	0.78
Poly(vinylidene fluoride)	0.8
Fluorinated ethylene--co-propylene	1.1
Poly(N, N'-$(p,p'$-oxydiphenylene) pyromellitide (Kapton)	1.1
Polystyrene	1.5–2.0
Polycarbonate	2.7
Poly(ethylene terephthalate)	3.6
Polypropylene	5.5
Low density polyethylene (LDPE)	6.0
High density polyethylene (HDPE)	6.2
Polyamide (Nylon 6)	7.2
Poly(methyl methacrylate)	22.6

Table 9.7. Activation energy for atomic oxygen exposed polymers [798]

Polymer	Activation energy $\kappa J\ mol^{-1}$
Polytetrafluoroethylene	1.0
Fluorinated ethylene-*co*-propylene	3.0
Poly(vinylidene fluoride)	6.2
Polystyrene	8.4
Poly(ethylene terephthalate)	9.8 ± 1.5
Poly(vinyl fluoride)	10.0
Poly(N,N'-(p,p-oxydiphenylene) pyromellitide (Kapton)	14.0
High density polyethylene (HDPE)	14.9
Low density polyethylene (LDPE)	14.1
Polycarbonate	15.4
Polyamide (Nylon 6)	19.1
Polypropylene	28.9
Poly(methyl methacrylate)	21.1

Atomic oxygen-induced mass loss is a thermally activated process. A summary of the activation energies are shown in Table 9.7. These activation energy values are smaller by a greater order of magnitude than the covalent bond energies [798].

Table 9.8 contains summaries of the effects of crosslinking, molecular weight, branching, fluorination, and bond strengths reflected in the mass loss equation τ

$$\tau = \phi^{-E_a/R_g T} \ .\tag{9.31}$$

where ϕ is the constrain, E_a is the activation energy, R_g is the gas constant and T is the absolute temperature (K)

Generation of the atomic oxygen in O_2 discharge reactors is always accompanied by other oxygen species such as $O(^1D)$, O_2^+, O_2^-, O^+, O^-, singlet oxygen $^1O_2(^1\Delta_g$ and $^1\Sigma_g^+)$, and free electrons. There are different methods to eliminate some of these oxygen species in order to obtain $O(^3D)$ and singlet oxygen $^1O_2$. However, laboratory conditions can never replace the space environment. In Table 9.9 comparative results of etch rates for various polymer films exposed to atomic oxygen in laboratory reactors and in the Space Shuttle STS-8 flight ($\approx$225 km above Earth) are presented.

9.14 Polymeric Photoinitiators Operating with a Fragmentation Mechanism

Polymeric photoinitiators are those polymers bearing photoreactive groups which, upon proper UV (light) irradiation, give rise to active species able to initiate the polymerization and crosslinking of mono- and poly-functional monomers and oligomers [128, 129].

From a technological point of view, polymeric photoinitiators operating with fragmentation (degradation) mechanisms are particularly convenient, due to their availability, high quantum yields of photodecomposition and high

Table 9.8. Material effects on mass loss equation parameters [798]

Effect	Polymer	Mass Loss Equation Parameters		
		Constraints (ϕ)	Amount of Mass Loss per Atomic Oxygen Atom (τ)	Activation(E_a) Energy
Crosslinking	Epoxies, polyamide Polyamide (Nylon 6), HDPE	Increase with cure	Constant	Increase with cure
		Increase with irradiation	Decrease with irradiation	Increase with irradiation
Chain scission	Kapton	Decrease with irradiation	Increase with irradiation	Decrease with irradiation
Molecular weight	Polyethylene, poly-styrene, poly (methyl methacrylate)	Increase with mol.wt.	Constant to mixed	Increase with mol.wt
Fluorination	Polytetra-fluoroethylene, fluorinated ethylene-co-propylene, poly(vinyl fluoride), poly (vinylidene fluoride)	Decrease with increasing fluorine	Decrease with increasing fluorine	Decrease with increasing fluorine
Branching	Polyethylene, polypropylene, polystyrene, (methyl methacrylate)	Relative to polyethylene (PE) decrease with ring , structure increase with methyl, longside groups	Relative to PE-decrease with methyl ring structure, increase with long side groups	Relative to PE-decrease with ring structure increase with methyl, longside groups

Table 9.9. Comparative etch rates for various polymer films exposed to atomic oxygen [300]

Polymer	O_2 plasma "in glow" (mg cm^{-2}/h)				Space Shuttle STS-8 Flight 225 km above Earth surface) (cm^3/ atom $\times$ 10^{24}) [438, 439]
	[331]	[437]	[763]	[300]	
Poly (N,N'-(p,p'-oxydiphenylene) pyromellitide (Kapton)	1.48	1.66	1.08	0.00155	3.0
Poly(ethylene terephthalate)	2.26	2.88	–	0.00232	3.4
Polystyrene	1.56	–	–	0.0298	1.7
Low density polyethylene (LDPE)	3.07	–	–	0.137	–
High density polyethylene (HDPE)	3.83	–	–	0.111	3.7
Poly(vinyl fluoride)	3.15	–	–	0.342	3.2
Poly(vinylidene fluoride)	–	–	–	0.0016	–
Polytetra-fluoroethylene	0.77	0.36	0.97	$\approx 3.7 \times 10^{-5}$	< 0.05

initiating efficiencies of the primary radicals in promoting the polymerisation of unsaturated monomers. A great number of UV (light) absorbing photo-initiating groups can be attached to the macromolecules which, under irradiation are photofragmentated.

(i) Copolymers of methyl methacrylate with 2,3-butanedione-2-O-metha-cryloyloxime $(R=CH_3)$ or 1-phenyl-1,2-propanedione-2-O-methacryloyloxime $(R=C_6H_5)$ (used for photografting of acrylamide or styrene are photofragmentated by the following mechanism [197, 704]:

$$
\begin{array}{c}
\text{CH}_3 \\
| \\
-\text{CH}_2-\text{C}- \\
| \\
\text{CO} \\
| \\
\text{O} \\
| \\
\text{O} \quad \text{N} \\
|| \quad || \\
\text{R}-\text{C}-\text{C}-\text{CH}_3 \\
\text{R}= \text{CH}_3, \text{C}_6\text{H}_5
\end{array}
\xrightarrow{\text{hv}}
\begin{array}{c}
\text{CH}_3 \\
| \\
-\text{CH}_2-\text{C}- \\
| \\
\text{CO} \\
| \\
\text{O} \\
\cdot
\end{array}
+ R-\overset{\cdot}{\text{CO}} + \text{CH}_3\text{CN}
$$

\qquad (9.32)

$$\downarrow M$$

homopolymer \qquad (9.33)

$$\downarrow M$$

graft-copolymer \qquad (9.34)

where M is the monomer.

When the above copolymers are irradiated as thin films at a temperature below their glass transition temperatures (T_g), they do not display any chain scission, $\overline{M}_n$ remaining practically unchanged [704]. It has been proposed that free-radical recombination in cage largely prevails below T_g; accordingly, if the films are plasticized so as to decrease their T_g below the irradiation temperature, again degradation occurs.

(ii) Polymeric systems bearing side-chain benzoin moieties are photofragmentated by the following mechanism [5, 47, 345, 423]:

$$
\begin{array}{c}
\text{R} \\
| \\
-\text{CH}_2-\text{CH}- \\
| \\
\text{CO} \\
| \\
\text{O} \quad \text{O} \\
|| \quad | \\
\langle\text{O}\rangle-\text{C}-\text{CH}-\langle\text{O}\rangle \\
\text{R}= \text{H}, \text{CH}_3
\end{array}
\xrightarrow{\text{hv}}
\begin{array}{c}
\text{R} \\
| \\
-\text{CH}_2-\text{CH}- \\
| \\
\text{CO} \\
| \\
\text{O} \\
| \\
\text{CH}-\langle\text{O}\rangle \\
\cdot
\end{array}
+ \langle\text{O}\rangle-\overset{\overset{\text{O}}{||}}{\text{C}} \cdot
$$

\qquad (9.35)

This photoinitiator, compared with a low-molecular-weight analogue in the UV-initiated polymerization of styrene, exhibited a markedly enhanced photocrosslinking efficiency.

(iii) The photo-α-cleavage polymers bearing 2-hydroxy-2-methylpropiophe-
none groups occurs by the mechanism [448, 449]:
The relative overall quantum yield of polymerization intiation (ϕ_i) is
expressed [244] by

(9.36)

$$\phi_i = \phi_\alpha \cdot \phi_{rm} \tag{9.37}$$

where ϕ_α and ϕ_{rm} are quantum yields of the α-photocleavage and initiating radicals formation, respectively.

(iv) Polymeric photointiators containing side chain benzophenone chromo-
phere [127, 129] under UV irradiation abstract hydrogen atoms and
produce polymeric ketyl radicals. In the presence of low molecular weight
amines or as a hybrid copolymer with amines, they produce active radicals which may intiate polymerization reactions. However, this process is
not based on a photo-fragmentation and will not be discussed further.

(v) Copolyesters derived from bisphenol-a and various amounts of 1-(4'-
hydroxyphenyl)-1, 2-propanedione-2-oxime or p-hydroxyphenylglyoxal
aldoxime and different diacid chlorides, consisting of repeating units

R_1=H, R_2= $+CH_2+_8$, m- or o-phenylene

gives upon irradiation block copolymers in the presence of vinyl or acrylic
monomers [197].

(vi) Block copolymers deriving from two different monomers can be prepared
by using a heterofunctional polymeric photoinitiator containing two
photolabile (azo- and peroxidic-) groups which differ sufficiently in selective absorption of the radiation [704]:

$$\left(\!\!-\overset{\displaystyle |}{\underset{\displaystyle O}{C}}\!\!-(CH_2)_2\!\!-\overset{\displaystyle CH_3}{\underset{\displaystyle CN}{C}}\!\!-N\!\!=\!\!N\!\!-\overset{\displaystyle CH_3}{\underset{\displaystyle CN}{C}}\!\!-(CH_2)_2\!\!-\overset{\displaystyle |}{\underset{\displaystyle O}{C}}\!\!-O\!\!-O\!\!-\!\!\right)_{\!\!n} \quad .$$

The selective photolysis of the azo groups at 350 and 371 nm in the presence of styrene leads to a styrene prepolymer containing acyl peroxidic groups which can be used further for polymerization of vinyl chloride.

More information on polymeric photointiators can be found elsewhere [128, 129, 175].

9.15   Positive-Working Photoresists

Photodegradation processes can be utilized in positive (deep-UV) resists, which are based on the fragmentation of the polymer chain, i.e. a reduction in polymer molecular weight, thereby increasing solubility in a suitable developer solvent enabling it to be washed out (Fig. 9.13).

A reasonable criterion for image discrimination is the condition that the molecular weight distributions of the exposed and unexposed polymer do not significantly overlap. In that case, any polymer molecule in the irradiated areas will dissolve faster than any molecule in the nonirradiated areas. Energy required to achieve this chain scission energy (E_r) is given by

$$E_r = wN_A 100/M_0 \phi_s f_s \tag{9.38}$$

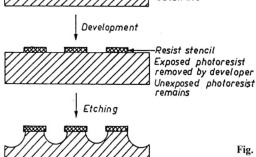

Fig. 9.13. Schematic representation of positive working photoresist

where w is the weight in grams of the irradiated sample

$$w = rd, \tag{9.39}$$

r is the sample thickness and d is the sample density, N_A is Avogadro's number, M_0 is the molecular weight of the repeat unit – the number of molecules of monomer units in the sample is wN_A/M_0, ϕ_s is the quantum yield for the main chain scission, and f_s is the fraction of bonds broken in the sample, given [135] by

$$f_s = \left(\frac{\overline{M}_{n(0)}}{\overline{M}_{n(t)}} - 1 \right) \frac{M_0}{\overline{M}_{n(0)}} \tag{9.40}$$

where $\overline{M}_{n(0)}$ and $\overline{M}_{n(t)}$ are initial and after bonds broken number-average molecular weights, respectively.

The sensitivity of an imaging system is determined by the amount of "detectable" change that is brought about by a single quantum of absorbed radiation.

A number of typical polymers and copolymers can be utilized as positive deep-UV resists [647]. The low sensitivity of poly(methyl methacrylate) to deep UV is caused by the low absorption coefficient of the polymer in this region. It requires $3400 \, \text{mJ cm}^{-2}$ (at 240 nm) to make poly(methyl methacrylate) developable. Nevertheless, the physical properties of poly(methyl methacrylate) and its high resolution capability are outstanding. The deficiency of its absorption can be removed by incorporating photo-sensitive (light absorbing) groups to the macromolecule. The poly(fluorobutyl methacrylate) has a sensitivity of $480 \, \text{mJ cm}^{-2}$ (at 240 nm) [502] and glycidyl-co-methyl methacrylate has a sensitivity of $250 \, \text{mJ cm}^{-2}$ (at 250 nm) [134].

$$\left[\begin{array}{c} \text{CH}_3 \\ | \\ \text{CH}_2\text{—C} \\ | \\ \text{CO} \\ | \\ \text{OC}_4\text{F}_9 \end{array} \right]_n$$

$$\left[\begin{array}{c} \text{CH}_3 \\ | \\ \text{CH}_2\text{—C} \\ | \\ \text{CO} \\ | \\ \text{O—CH}_2\text{—CH—CH}_2 \\ \diagdown \diagup \\ \text{O} \end{array} \right]_n \left[\begin{array}{c} \text{CH}_3 \\ | \\ \text{CH}_2\text{—C} \\ | \\ \text{CO} \\ | \\ \text{O—CH}_3 \end{array} \right]_m$$

Incorporation of acyloximino groups into poly (methyl methacrylate), e.g. 3-oximino-2-butanone methacrylate-*co*-methyl methacrylate, causes this co-polymer to have a strong absorption around 225 nm and it is highly reactive [800]. The fragmentation of the copolymer chain occurs by the following mechanism:

$$
\begin{array}{c}
\quad\;\; CH_3 \qquad\qquad\qquad CH_3 \\
\quad\;\; | \qquad\qquad\qquad\qquad | \\
-CH_2-C\!\!-\!\!-\!\!-\!\!-\!\!-\!\!-\!\!-\!\!-\!\!CH_2-C- \quad\xrightarrow{\; h\nu \;} \\
\quad\;\; | \qquad\qquad\qquad\qquad | \\
\quad\;\; CO \qquad\qquad\qquad\;\; CO \\
\quad\;\; | \qquad\qquad\qquad\qquad | \\
\quad\;\; O \qquad\qquad\qquad\;\; O-CH_3 \\
\quad\;\; | \\
\quad\;\; N\!\!=\!\!C-CO-CH_3 \\
\qquad\;\; | \\
\qquad\;\; CH_3
\end{array}
$$

$$
\begin{array}{c}
\quad\;\; CH_3 \qquad\qquad\qquad CH_3 \\
\quad\;\; | \qquad\qquad\qquad\qquad | \\
-CH_2-C\!\!-\!\!-\!\!-\!\!-\!\!-\!\!-\!\!-\!\!-\!\!CH_2-C- \quad + \; CH_3CO\cdot + CH_3CN \longrightarrow \\
\quad\;\; | \qquad\qquad\qquad\qquad | \\
\quad\;\; CO \qquad\qquad\qquad\;\; CO \\
\quad\;\; | \qquad\qquad\qquad\qquad | \\
\quad\;\; O \qquad\qquad\qquad\;\; O-CH_3 \\
\quad\;\; \cdot
\end{array}
$$

$$
\begin{array}{c}
\quad\;\; CH_3 \qquad\qquad\qquad CH_3 \\
\quad\;\; | \qquad\qquad\qquad\qquad | \\
-CH_2-C\!\!-\!\!-\!\!-\!\!-\!\!-\!\!-\!\!-\!\!-\!\!CH_2-C- \quad + CO_2 \; \xrightarrow{\; \beta\text{-scission} \;} \\
\quad\;\; \cdot \qquad\qquad\qquad\qquad | \\
\qquad\qquad\qquad\qquad\qquad CO \\
\qquad\qquad\qquad\qquad\qquad | \\
\qquad\qquad\qquad\qquad\qquad O-CH_3
\end{array}
$$

$$
\begin{array}{c}
\quad\;\; CH_3 \qquad\qquad CH_3 \\
\quad\;\; | \qquad\qquad\quad\; | \\
-CH_2\!\!=\!\!C \quad + \cdot CH_2-C- \\
\qquad\qquad\qquad\qquad | \\
\qquad\qquad\qquad\qquad CO \\
\qquad\qquad\qquad\qquad | \qquad\qquad\qquad\qquad\qquad (9.41)\\
\qquad\qquad\qquad\qquad O-CH_3
\end{array}
$$

Incorporation of 16 mol% of acyloximino groups enhances copolymer sensitivity to 80 mJ cm^{-1} (at 240 nm). The sensitivity of the system can be further increased by preparing a terpolymer containing methylacrylonitrile as the third component. The most photo-sensitive polymer contains these three components in the ratio 15:70:15:

This terpolymer has a sensitivity of $40\,mJ\,cm^{-2}$ (at 240 nm), which can be further enhanced by a sensitization process using *p-tert*-butylbenzoic acid as a sensitizer [645].

Another high sensitivity positive-working photoresist was obtained by the copolymerization of methyl methacrylate with indenone [646]. This copolymer has a strong absorption in the 230 to 300 nm region and a sensitivity of $20\,mJ\,cm^{-2}$. The fragmentation of the co-polymer chain occurs by the following mechanism:

$$(9.42)$$

It has been proposed as a positive working photoresist based on the Norrish Type I photocleavage of poly (methylisopropenyl ketone) [737, 773]:

$$-CH_2-\underset{\underset{\displaystyle CH_3}{\overset{\displaystyle CO}{|}}}{\overset{\displaystyle CH_3}{\underset{|}{\overset{|}{C}}}}-CH_2-\underset{\underset{\displaystyle CH_3}{\overset{\displaystyle CO}{|}}}{\overset{\displaystyle CH_3}{\underset{|}{\overset{|}{C}}}}-\xrightarrow{h\nu}-CH_2-\overset{\displaystyle CH_3}{\underset{\displaystyle \cdot}{\underset{|}{\overset{|}{C}}}}-CH_2-\underset{\underset{\displaystyle CH_3}{\overset{\displaystyle CO-CH_3}{|}}}{\overset{\displaystyle CH_3}{\underset{|}{\overset{|}{C}}}}- +CO+\cdot CH_3$$

β-chain scission

$$-CH_2-\overset{\displaystyle CH_3}{\underset{|}{\overset{|}{C}}}=CH_2 +\cdot\underset{\underset{\displaystyle CH_3}{\overset{\displaystyle CO}{|}}}{\overset{\displaystyle CH_3}{\underset{|}{\overset{|}{C}}}}-$$

$$(9.43)$$

The sensitivity of this polyketone is $700\,\mathrm{mJ\,cm^{-2}}$ (at 280 nm).

An interesting group of positive working photoresists is based on polymers containing the triphenylsulphonium ion in their backbone [154]. Triphenylsulphonium salts and other onium salts act as photoinitiators in cationic polymerization.

On irradiation they undergo facile bond cleavage:

$$Ph_3S^+X^- \xrightarrow{h\nu} Ph\cdot + Ph_2S^+X^- \; . \tag{9.44}$$

Polyimides which have the triphenylsulphonium moiety built into the backbone are photo-fragmented by the following mechanisms [154]:

$$(9.45)$$

$$(9.46)$$

The technological application of positive-working resist is in photolithography. After exposure to a light source that registers a pattern, the photoresist is developed by wet chemical methods [9].

10 References

1. Adam W (1974) Chem Commun 289
2. Adams JH (1970) J Polym Sci 8:1077
3. Adams MR, Garton A (1993) Polym Degrad Sci 41:265
4. Ahmad SR (1983) J Phys Appl Phys 16:L137
5. Ahn KD, Ihn KJ, Kwon IC (1986) J Macromol Sci Chem A23:355
6. Akaishi D, Tsubomura SH, Yamamoto N (1972) Kobunshi Kagaku 29:657
7. Akay G, Tincer T (1981) Polym Eng Sci 21:8
8. Akay G, Tincer T, Ergöz HE (1980) Europ Polym J 16:601
9. Allen DM (1986) The Principles and Practice of Photochemical Machining and Photo-etching, Hilger, Bristol
10. Allen NS (1981) Dev Polym Degrad 2:239
11. Allen NS (1986) Chem Soc Rev 15:373
12. Allen NS (1987) Rev Prog Color 17:61
13. Allen NS (1989) In: Comprehensive Polymer Science (Allen G, Bevington JC, eds) vol 6, p 579
14. Allen NS (1994) Polym Degrad Stabil 44:357
15. Allen NS, Edge M (1992) Fundamentals of Polymer Degradation and Stabilization, Elsevier Applied Science, London
16. Allen NS, Edge M, Mohammadian M, Jones K (1994) Polym Degrad Stabil 43:229
17. Allen NS, Fatinikum KO (1980/81) Polym Degrad Stabil 3:327
18. Allen NS, Fatinikum KO, Gardette J, Lemaire J (1982) Polym Degrad Stabil 4:95
19. Allen NS, Fatinikum KO, Henman TJ (1982) Polym Degrad Stabil 4:59
20. Allen NS, Fatinikum KO, Henman TJ (1983) Europ Polym J 19:551
21. Allen NS, Harrison MJ, Follows GW, Matthews V (1987) Polym Degrad Stabil 19:77
22. Allen NS, Homer J, McKellar JF (1976) J Appl Polym Sci 20:2553
23. Allen NS, Homer J, McKellar JF (1977) J Appl Polym Sci 21:2261
24. Allen NS, Homer J, McKellar JF, Wood GDM (1977) J Appl Polym Sci 21:3147
25. Allen NS, Marshall GP, Vasiliou C, Moore LM, Kotecha JL, Gardette JL (1988) Polym Degrad Stabil 20:315
26. Allen NS, McKellar JF (1977) Brit Polym J 9:302
27. Allen NS, McKellar JF (1977) Polymer 18:968
28. Allen NS, McKellar JF (1978) J Appl Polym Sci 22:625
29. Allen NS, McKellar JF (1979) Makromol Chem 180:2875
30. Allen NS, McKellar JF (1979) Dev Polym Degrad 2:129
31. Allen NS, McKellar JF (1980) In: Photochemistry of Dyed and Pigmented Polymers (Allen NS, McKellar JF, eds) Applied Science Publishers, London, p 247
32. Allen NS, McKellar JF, Phillips GO (1975) J Polym Sci Chem Ed 13:2857
33. Allen NS, McKellar JF, Phillips GO (1977) Am Chem Soc Polym Preprints 18:375
34. Allen NS, McKellar JF, Protopapas SA (1978) J Appl Polym Sci 22:1451
35. Allen NS, McKellar JF, Wilson D (1976/77) J Photochem 6:337
36. Allen NS, McKellar JF, Wilson D (1977) J Polym Sci Chem Ed 15:2793
37. Allen NS, Wilson D, McKellar JF (1978) Makromol Chem 179:269
38. Al-Malaika S (1989) In: Comprehensive Polymer Science (Allen G, Bevington JC, eds) Pergamon Press, Oxford, p 539
39. Al-Malaika S, Chakraborthy KN, Scott G (1983) Dev Polym Stabil 6:73
40. Al-Malaika S, Marogi A, Scott G (1985) J Appl Polym Sci 30:789
41. Al-Malaika S, Marogi A, Scott G (1986) J Appl Polym Sci 31:685
42. Al-Malaika S, Marogi A, Scott G (1987) J Appl Polym Sci 33:1455
43. Amerik Y, Guillet JE (1971) Macromolecules 4:375
44. Amin MU, Scott G (1974) Europ Polym J 10:1019

45. Andrady AL (1990) J Appl Polym Sci 39:363
46. Andrew JE, Dyer PE, Forster D, Key PH (1983) Appl Phys Lett 43:717
47. Angiolini L, Carlini C (1980) Chim Ind 72:124
48. Arnaud R, Lemaire J (1979) Dev Polym Degrad 2:159
49. Arnaud R, Lemaire J (1981) Dev Polym Photochem 2:135
50. Arnold BR, Scaiano JC (1992) Macromolecules 25:1582
51. Arnold GS, Peplinski DR (1985) AIAA J 23:1621
52. Asquith RS, Gardner KL, Geehan TG, McNally G (1977) J Polym Sci Polym Lett 15:435
53. Audonin L, Langlois V, Verdu J, de Bruijn JCM (1994) J Mater Sci 29:569
54. Augustyniak W, Wojtczak J (1980) J Polym Sci Chem Ed 18:1339
55. Bachmann F (1989) Chemotronics 4:149
56. Bachmann F (1989) Mater Res Soc Bull 14:49
57. Bahners T, Schollmeyer E (1987) Angew Makromol Chem 151:1 and 39
58. Bahners T, Schollmeyer E (1988) Makromol Chem Rapid Commun 9:115
59. Bahners T, Schollmeyer E (1989) J Appl Phys 66:1884
60. Balazs GH (1985) In: Proceedings of the Workshop on the Fate and Impact Marine Debris
    (Shomura ES, Yoshida HO, eds) Honolulu, Hawaii, p 387
61. Ballara A, Verdu J (1989) Polym Degrad Stabil 26:361
62. Barltrop JA, Coyle JD (1978) Principles of Photochemistry, Wiley,
    New York
63. Bartos J, Tino J (1984) Polymer 25:274
64. Basile LJ (1965) Trans Faraday Soc 42:3163
65. Baum B (1974) Polym Eng Sci 14:205
66. Baumhardt-Neto R, De Paoli MA (1993) Polym Degrad Stabil 40:53
67. Baumhardt-Neto R, De Paoli MA (1993) Polym Degrad Stabil 40:59
68. Beachell HC, Chang IL (1972) J Polym Sci Chem Ed 10:503
69. Bellenger V, Bouchard C, Claveirolle P, Verdu J (1981) Polym Photochem 1:69
70. Bellenger V, Verdu J (1983) J Appl Polym Sci 28:2677
71. Bellenger V, Verdu J (1984) Polym Photochem 5:295
72. Bellenger V, Verdu J (1985) J Appl Polym Sci 30:363
73. Bellenger V, Verdu J, Martines G, Millan J (1990) Polym Degrad Stabil 28:53
74. Belluš D (1971) Adv Photochem 8:109
75. Belluš D, Hrdlovič P, Maňásek Z (1966) J Polym Sci Polym Lett 4:1
76. Benachour D, Rogers CE (1983) ACS Symp Ser 220:307
77. Benson SW (1964) J Chem Phys 40:1007
78. Benson SW (1965) J Chem Educ 42:501
79. Bhateja SK, Andrews EH (1985) J Mater Sci 20:2839
80. Bigger SW, Delatycki O (1987) J Polym Sci Chem Ed 25:3311
81. Bigger SW, Delatycki O (1988) Polymer 29:1277
82. Billingham NC, Calvert PD (1980) Dev Polym Stabil 3:139
83. Billingham NC, Walker TJ (1975) J Polym Sci Chem Ed 13:1209
84. Birks JB (1970) Prog Reaction Kinet 5:181
85. Birks JB (1970) Photophysics of Aromatic Molecules, Wiley, New York
86. Blum SE, Brown KH, Srinivasan R (1983) US Patent 4,414,059
87. Bokobza L, Jasse B, Monnerie L (1980) Europ Polym J 16:715
88. Bolton JR, Archer MD (1991) ACS Symp Ser 228:7
89. Boss CR, Chien JCW (1966) J Polym Sci Chem Ed 4:1543
90. Bousquet JA, Fouassier JP (1982) J Photochem 20:197
91. Bousquet JA, Fouassier JP (1983) Polym Degrad Stabil 5:113
92. Bousquet JA, Fouassier JP (1984) Europ Polym J 20:983
93. Bousquet JA, Fouassier JP (1984) J Polym Sci Chem Ed 22:3865
94. Bousquet JA, Fouassier JP (1987) Angew Makromol Chem 149:19
95. Bousquet JA, Fouassier JP (1987) Angew Makromol Chem 149:45
96. Bousquet JA, Fouassier JP (1987) Polym Degrad Stabil 18:163
97. Bousquet JA, Fouassier JP (1987) Europ Polym J 23:367
98. Brannon JH (1989) J Vac Sci Technol B7:1064
99. Brannon JH, Lankard JR (1986) Appl Phys Lett 48:1226
100. Brannon JH, Lankard JR, Baise AI, Burns F, Kaufman J (1985)
     J Appl Polym Sci 58:2036
101. Braren B, Seeger D (1986) J Polym Sci Polym Lett 24:371
102. Braun H, Kovacs G (1963) J Phys Chem Glasses 4:152
103. Braun W, Rajbenbach L, Eirich FR (1962) J Phys Chem 66:1591

104. Brinza DE (ed) (1987) Proceedings of NASA Workshop on Atomic Oxygen Effects, JPL Publ. p 87
105. Brinza DE, Stiegman AE, Staszak PR, Laue EG, Liang RH (1991) Proceedings of NASA LDEF 69 Months in Space Symposium, NASA, Part 2, p 817
106. Brune WH, Stimpfle RM (1993) ACS Symp Ser 232:133
107. Bunce NJ (1984) J Photochem 15:1
108. Bunce NJ (1987) J Photochem 38:99
109. Buettner GR (1987) Free Radical Biol Med 3:259
110. Burkhart RD, Nam-In J (1993) ACS Symp Ser 236:599
111. Busfield WK, Monteiro J (1990) Mater Forum 14:218
112. Calvert JG, Pitts JN (1986) Photochemistry, Wiley, New York
113. Cameron GC, Bullock AT (1982) Dev Polym Charact 3:107
114. Cannon RD (1980) Electron Transfer Reactions, Butterworth, London
115. Carlsson DJ, Brousseau R, Zhang C, Wiles DM (1988) ACS Symp Ser 364:376
116. Carlsson DJ, Chan KH, Durmis J, Wiles D (1982) J Polym Sci Chem Ed 20:575
117. Carlsson DJ, Dobbin CJB, Wiles DM (1985) Macromolecules 18:1791
118. Carlsson DJ, Garton A, Wiles DM (1976) Macromolecules 9:695
119. Carlsson DJ, Garton A, Wiles DM (1979) Dev Polym Stabil 1:219
120. Carlsson DJ, Parnell RD, Wiles DM (1973) J Polym Sci Polym Lett 11:149
121. Carlsson DJ, Wiles DM (1969) Macromolecules 2:587
122. Carlsson DJ, Wiles DM (1969) Macromolecules 2:597
123. Carlsson DJ, Wiles DM (1971) Macromolecules 4:179
124. Carlsson DJ, Wiles DM (1974) J Polym Sci Chem Ed 12:2217
125. Carlsson DJ, Wiles DM (1974) Rubb Chem Technol 49:991
126. Carlsson DJ, Wiles DM (1976) J Macromol Sci Rev Macromol Chem C14:65
127. Carlini C (1986) Brit Polym J 18:236
128. Carlini C, Angiolini L (1993) In: Radiation Curing in Polymer Science and Technology (Fouassier JP, Rabek JF, eds) Elsevier Applied Science, London, vol II, p 283
129. Carlini C, Angiolini L (1995) Adv Polym Sci 123:128
130. Carraher CEJr (1983) ACS Symp Ser 229:25
131. Carroll WF, Schissel P (1980) Polymers in Solar Technologies: An R&D Strategy, Rport SERI/TR-334-601, Solar Energy Research Institute, Golden, Ohio
132. Carroll WF, Schissel P (1983) ACS Symp Ser 220:3
133. Casale A, Porter RS (1978) Polymer Stress Reaction, Academic Press, New York
134. Chandross EA, Reichmanis E, Wilkins CW, Hartless RL (1981) Solid State Technol 24:81
135. Charlesby A (1960) Atomic Radiation and Polymers, Pergamon Press, Oxford
136. Charlesby B, Grace CS, Pilkington FB (1962) Proc Roy Soc A268:205
137. Chatgilialoglu C, Ingold KU (1981) J Amer Chem Soc 103:4833
138. Chew CH, Gan LM, Scott G (1977) Europ Polym J 13:361
139. Chien JCW, Boss CR (1967) J Polym Sci Chem Ed 5:3091
140. Chien JCW, Jabloner H (1968) J Polym Sci Chem Ed 6:393
141. Chien JCW, Vandenberg EJ, Jabloner H (1968) J Polym Sci Chem Ed 6:381
142. Chipalkatti MH, Laski JJ (1993) ACS Symp Ser 236:262
143. Christ LC, Gregory JC, Peters PN (1991) Proceedings of NASA LDEF 69 Months in Space Symposium, NASA, Part 2, p 723
144. Chryssolouris G (1991) Laser Machining:Theory and Practice, Springer, Berlin
145. Clark DT, Munro HS (1984) Polym Degrad Stabil 8:213
146. Clark DT, Munro HS (1984) Polym Degrad Stabil 9:63
147. Clark LG, Kinard WH, Carter DJ, Jones JL (eds) (1984) The Long Duration Exposure Facility (LDEF), NASA, SP-473
148. Clough RL, Dillon MP, Iu KK, Ogilby PR (1989) Macromolecules 22:3620
149. Cozzens RF, Fox RB (1969) J Chem Phys 50:1532
150. Cozzens RF, Fox RB (1978) Polym Eng Sci 18:900
151. Crank J (1975) The Mathematics of Diffusion, 2nd ed, Oxford University Press, Oxford
152. Crank J, Park GS (1968) Diffusion in Polymers, Academic Press, New York
153. Creed D, Holye CH, Subramanian P, Nagarajan R, Pandey C, Anzures ET, Cane KM, Cassidy PE (1994) Macromolecules 27:832
154. Crivello JV, Lee JL, Conlon DA (1987) J Polym Sci Chem Ed 25:3293
155. Cunliffe AV, Davis A (1982) Polym Degrad Stabil 4:17
156. Curlee TR, Das S (1986) Plastic Wastes, Noyes Data Co, Park Ridge, New Jersey
157. Curro JG, Lagasse RR, Simha R (1982) Macromolecules 15:1621

158. Czekajewski J, Nennerfelt L, Kaczmarek H, Rabek JF (1994) Acta Polym 45:369
159. Dan E, Guillet JE (1973) Macromolecules 6:230
160. David C (1975) In: Degradation of Polymers (Bamford CH, Tipper CFH, eds) Compre-
     hensive Chemical Kinetics, Elsevier, Amsterdam, vol 14, p 175
161. David C, Bayens-Volant D (1978) Europ Polym J 14:29
162. David C, Borsu M, Geuskens G (1970) Europ Polym J 6:959
163. David C, Camara O, Geuskens G (1977) Polymer 18, 198
164. David C, Demarteau W, Geuskens G (1967) Polymer 8:497
165. David C, Demarteau W, Geuskens G (1970) Europ Polym J 6:537
166. David C, Demarteau W, Geuskens G (1970) Europ Polym J 6:1405
167. David C, Demarteau W, Geuskens G (1971) Europ Polym J 6:1397
168. David C, Lempereur M, Geuskens G (1972) Europ Polym J 8:417
169. David C, Naegelen V, Piret W, Geuskens G (1975) Europ Polym J 11:569
170. David C, Putman N, Lempereur M, Geuskens G (1972) Europ Polym J 8:409
171. David C, Trojan M, Daro A, Demarteau W (1992) Polym Degrad Stab 37:233
172. David C, Trojan M, Daro A, Weiland M (1990) Polym Mater Sci Eng 63:951
173. Davidson JA, Abrahamson EW (1972) Photochem Photobiol 15:403
174. Davidson JA, Kear KE, Abrahamson EW (1972/73) J Photochem 1:297
175. Davidson RS (1993) J Photochem Photobiol A Chem 69:263
176. Davis A, Sims D (1983) Weathering of Polymers, Applied Science Publishers, New York
177. Debye PJ (1942) Trans Electrochem Soc 82:265
178. Decker C (1976) J Appl Polym Sci 20:3321
179. Decker C (1984) Europ Polym J 20:149
180. Decker C (1984) In: Degradation and Stabilization of Poly(vinyl chloride), Elsevier, Lon-
     don, p 81
181. Decker C (1987) J Polym Sci Polym Lett 25:5
182. Decker C (1988) ACS Symp Ser 364:201
183. Decker C (1989) Makromol Chem Makromol Symp 24:253
184. Decker C (1992) Makromol Chem Makromol Symp 57:103
185. Decker C, Ackhard A, Ehrburger P (1990) Carbon 28:246
186. Decker C, Balandier M (1980) Makromol Chem Rapid Commun 1:389
187. Decker C, Balandier M (1981) Polym Photochem 1:221
188. Decker C, Balandier M (1981) J Photochem 15:213
189. Decker C, Balandier M (1981) J Photochem 15:221
190. Decker C, Balandier M (1982) Europ Polym J 18:1085
191. Decker C, Balandier M (1982) Makromol Chem 183:1263
192. Decker C, Balandier M (1984) Polym Photochem 5:267
193. Decker C, Mayo FR, Richardson H (1973) J Polym Sci Chem Ed 11:2879
194. DeGennes PG (1971) J Chem Phys 55:572
195. DeGennes PG (1982) J Chem Phys 76:3316
196. Delprat P, Gardette JL (1993) Polymer 34:933
197. Delzenne GA, Laridon U, Peeters H (1970) Europ Polym J 6:933
198. Denisov ET (1982) Dev Polym Degrad 5:23
199. Deutsch TF, Geis MW (1983) J Appl Phys 54:7201
200. Dexter DL (1953) J Chem Phys.(1953) 21:836
201. Dickie A, Vries JS, Holubka JV (1982) Anal Chem 54:2045
202. Dobrowolski DC, Ogilby PR, Foote CS (1983) J Phys Chem 87:2261
203. Dole M, Salik J (1977) J Amer Chem Soc 99:6454
204. Donald AM (1985) J Mater Sci 20:2630
205. Dunderdale J (ed) (1990) Energy and the Environment, Royal Society of Chemistry, London
206. Dunn DS, Ouderkirk AJ (1990) Macromolecules 23:770
207. Durán N, Gómez E, Mansilla H (1987) Polym Degrad Stabil 17:131
208. Dyer PE (1992) In: Photochemical Processing of Electronic Materials, Academic Press, New
     York, p 359
209. Dyer PE, Farley RJ (1990) Appl Phys Lett 57:765
210. Dyer PE, Jenkins SD, Sidhu J (1988) Appl Phys Lett 52:1880
211. Dyer PE, Oldershaw GA, Sidhu J (1991) J Phys Chem 95:1004
212. Dyer PE, Sidhu J (1985) J Appl Phys 57:1420
213. Dyer PE, Sidhu J (1986) J Opt Soc Am 3:792
214. Dyer PE, Srinivasan R (1986) J Appl Phys Lett 48:445
215. Egerton GS (1971) Brit Polym J 3:63
216. Egerton GS, Morgan AG (1971) J Soc Dyers Colour 87:208

217. Ehrlich DJ, Tsao JY, Bozler CO (1985) J Vac Sci Technol B3:1
218. El-Sayed M (1963) J Chem Phys 38:2834
√ 219. Emanuel NM, Roginsky BA, Buchachenko AL (1982) Uspekh Khim 51:361
220. Encinas MV, Funabashi K, Scaiano JC (1979) Macromolecules 12:1167
221. Encinas MV, Funabashi K, Scaiano JC (1984) Macromolecules 17:2261
222. Erben F, Vesely R (1983) Angew Makromol Chem 200:161
223. Ermolaev VL (1968) Bull Nat Acad Sci USSR 32:1193
224. Everhart S, Reiley CN (1981) Anal Chem 53:665
225. Evleth EM (1976) J Amer Chem Soc 98:1637
226. Factor A, Chu ML (1980) Polym Degrad Stabil 2:203
227. Factor A, Ligon MV, May RJ (1987) Macromolecules 20:2461
228. Factor A, Lynch JC, Greenberg FH (1987) J Polym Sci Polym Ed 25:3413
229. Fairgrieve SP, MacCallum JR (1985) Polym Degrad Stabil 11:251
√230. Falicki S, Carlsson DJ (1993) Polym Degrad Stabil 41:205
231. Faure J, Fouassier JP, Lougnot DJ (1977) Nouv J Chim 1:15
232. Felder B (1985) ACS Symp Ser 280:69
233. Ferguson DC (1984) NASA Report CP-2340
234. Fewell LL, Finney L (1991) Polym Commun 32:393
235. Finlayson-Pitts BJ, Pitts JNJr (1986) Atmospheric Chemistry: Fundamentals and Experi-
     mental Technique, Wiley, New York
236. Fitzgibbon PD, Frank CW (1982) Macromolecules 15:733
237. Förster T (1946) Naturwiss 33:166
238. Förster T (1953) Disc Faraday Soc 21:836
239. Förster T (1959) Disc Faraday Soc 27:7
240. Fossey J, Lefort D, Sorba J (1993) Topics Curr Chem 164:100
241. Fouassier JP (1989) In: Photochemistry and Photophysics (Rabek JF, ed) CRC Press, Boca
     Raton, vol II, p 1
242. Fouassier JP (1993) In: Radiation Curing in Polymer Science and Technology (Fouassier JP,
     Rabek JF eds) Elsevier Applied Science, London, vol I, p 49
243. Fouassier JP (1993) In: Radiation Curing in Polymer Science and Technology (Fouassier JP,
     Rabek JF, eds) Elsevier Applied Science, London, vol II, p 1
244. Fouassier JP, Lougnot DJ, Li Bassi G, Nicora C (1989) Polym Commun 30:245
245. Fox RB (1972) Pure Appl Chem 30:87
246. Fox RB (1973) Pure Appl Chem 34:235
247. Fox RB, Isaacs LG, Stokes S (1963) J Polym Sci Chem Ed 1:1079
248. Francois J, Candau F, Benoit H (1974) Polymer 15:618
249. Frank CW, Harrah IA (1974) J Chem Phys 61:1526
250. Fredrickson GH, Andersen HC, Frank CW (1983) J Chem Phys 79:3572
251. Fredrickson GH, Anderson HC, Frank CW (1983) Macromolecules 16:3572
252. Fredrickson GH, Anderson HC, Frank CW (1984) Macromolecules 17:54
253. Fredrickson GH, Anderson HC, Frank CW (1984) Macromolecules 17:1496
254. Fredrickson GH, Frank CW (1983) Macromolecules 16:572
255. Fredrickson GH, Frank CW (1983) Macromolecules 16:1198
256. Friedrich K (1983) Adv Polym Sci 52/53:1
257. Funke W, Haagen H (1983) ACS Symp Ser 229:309
258. Furneaux GC, Ledbury KJ, Davis A (1980) Polym Degrad Stabil 3:431
259. Gardette JL, Gaumet S, Philipart JL (1993) J Appl Polym Sci 48:1885
260. Gardette JL, Lemaire J (1981) Makromol Chem 182:2723
261. Gardette JL, Lemaire J (1982) Makromol Chem 183:2415
262. Gardette JL, Lemaire J (1984) Polym Degrad Stabil 6:135
263. Gardette JL, Lemaire J (1986) Polym Degrad Stabil 16:147
264. Gardette JL, Lemaire J (1987) Rev Gen Caoutch Plast 672:133
265. Gardette JL, Lemaire J (1989) Polym Degrad Stabil 25:293
266. Gardette JL, Lemaire J (1991) Polym Degrad Stabil 33:77
267. Gardette JL, Philippart JL (1988) J Photochem Photobiol A Chem 43:221
268. Garner BL, Papello PJ (1962) Ind Eng Chem 1:249
269. Garrison BJ, Srinivasan R (1984) Appl Phys Lett 44:849
√ 270. Garton A, Carlsson DJ, Wiles DM (1978) J Polym Sci Chem Ed 16:33
√271. Garton A, Carlsson DJ, Wiles DM (1979) Macromolecules 12:1071
272. Garton A, Carlsson DJ, Wiles DM (1980) J Polym Sci Chem Ed 18:3245
273. Garton A, Carlsson DJ, Wiles DM (1980) Dev Polym Photochem 1:93
274. Garton A, McLean PD, Wiebe W, Densley RJ (1986) J Appl Polym Sci 32:3941

275. Geetha R, Torikai A, Nagaya S, Fueki K (1987) Polym Degrad Stabil 19: 279
276. George GA, Ghaemy M (1991) Polym Degrad Stabil 33:411
277. George GA, Hodgeman DKC (1976) J Polym Sci Symp 55:195
278. Geuskens G (ed) (1975) Degradation and Stabilization of Polymers, Wiley, New York
279. Geuskens G (1975) In: Degradation of Polymers (Bamford CH, Tipper CFH, eds) Comprehensive Chemical Kinetics, Elsevier Scientific Publishers Amsterdam, vol 14, p 333
280. Geuskens G, Baeyens-Volant D, Delaunois G, Lu-Vinh Q, Piret W, David C (1978) Europ Polym J 14:291
281. Geuskens G, Baeyens-Voland D, Delaunois G, Lu-Vinh Q, Piret W, David C (1978) Europ Polym J 14:299
282. Geuskens G, David C (1971) IUPAC Special Lecture 8:19
283. Geuskens G, David C (1977) Pure Appl Chem 49:479
284. Geuskens G, David C (1979) Pure Appl Chem 51:233
285. Geuskens G, David C (1979) Pure Appl Chem 51:2385
286. Geuskens G, Delaunois D, Lu-Vinh Q, Piret W, David C (1978) Europ Polym J 14:299
287. Geuskens G, Debie F, Kabamba M, Nedelkos G (1984) Polym Photochem 5:313
288. Geuskens G, Kabamba M (1987) Polym Degrad Stabil 19:315
289. Ghiggino KP, Barraclough CG, Harris RJ (1993) ACS Sym Ser 527:155
290. Ghiggino LP, Nicholls CH, Pailthorpe MT (1975) J Photochem 4:155
291. Gilead D (1985) Chemtech (May) 299
292. Gilead D (1990) Polym Degrad Stabil 29:65
293. Gilead D, Scott (1982) Dev Polym Stabil 5:71
294. Gillen KT, Clough RL (1991) ACS Symp Ser 475:457
295. Gillispie GD (1993) ACS Symp Ser 236:89
296. Giori C, Yamauchi T (1984) J Appl Polym Sci 29:237
297. Golub MA (1980) Pure Appl Chem 52:305
298. Golub MA, Rosenberg ML, Gemmer RV (1977) Rubber Chem Technol 50:704
299. Golub MA, Stephens CL (1966) J Polym Sci Polym Lett 4:959
300. Golub MA, Wydeven T (1988) Polym Degrad Sci 22:325
301. Graedel TE (1978) Chemical Compounds in the Atmosphere, Academic Press, New York
302. Grassie N, Scott G (1985) Polymer Degradation and Stabilization, Cambridge University Press, Cambridge
303. Grassie N, Weir NA (1965) J Appl Polym Sci 9:963
304. Grassie N, Weir NA (1965) J Appl Polym Sci 9:975
305. Grattan DW, Carlsson DJ, Wiles DM (1978) Chem Ind 228
306. Gugumus F (1989) Makromol Chem Makromol Symp 27:25
√307. Gugumus F (1990) Polym Degrad Stabil 27:19
308. Guillet JE (1972) Pure Appl Chem 30:135
309. Guillet JE (1972) Naturwissensch 59:503
310. Guillet JE (1973) In: Polymer Science and Technology (Guillet JE, ed) Plenum Press, New York, vol 3, p 1
311. Guillet JE (ed) (1973) Polymers and Ecological Problems, Plenum, New York
312. Guillet JE (1974) Polym Eng Sci 14:482
313. Guillet JE (1974) Plast Eng 30:48
314. Guillet JE (1977) Pure Appl Chem 49:249
315. Guillet JE (1978) Adv Chem Ser 169:1
316. Guillet JE (1979) J Appl Polym Sci Appl Polym Symp 35:599
317. Guillet JE (1985) Polymer Photophysics and Photochemistry, Cambridge University Press, Cambridge
318. Guillet JE (1990) In: Degradable Materials (Barenberg SA, Brash JL, Narayan R, Redpath AE eds) CRC Press, Boca Raton, p 55
319. Guillet JE, Dhanraj J, Golemba FJ, Hartley GH (1968) Adv Chem Ser 85:272
320. Guillet JE, Li SKL, Ng HC (1984) ACS Symp Ser 266:165
321. Guillet JE, Norrish RGW (1955) Proc Roy Soc A 233:153
322. Gupta MC, Gupta A (1983) Polym Photochem 3:211
323. Gupta MC, Gupta A, Horowitz Y, Kilger D (1982) Macromolecules 15:1372
324. Gupta A, Rembaum A, Moacanin J (1978) Macromolecules 11:1285
325. Haag WR, Hoigne J (1985) Chemosphere 14:1659
326. Hama Y, Hoscono K, Furui Y, Shinohara K (1966) Rep Prog Polym Phys Japan 12:1433
327. Hama Y, Hoscono K, Furui Y, Shinohara K (1971) J Polym Sci Chem Ed 9:1411
328. Hamid SH, Amin MB, Maadhah AG (eds) Handbook of Polymer Degradation, Dekker, New York

329. Hansen RH (1968) In: Interface Conversion (Weiss P, Cheever GD, eds) Elsevier, New York, p 287
330. Hansen RH (1970) In: Thermal Stability of Polymers (Conley RT, ed) Dekker, New York, p 153
331. Hansen RH, Pascale JV, Debendictis T, Rentzepis PM (1965) J Polym Sci Chem Ed 3:2205
332. Harbour JR, Hair ML (1986) Adv Colloid Interface Sci 24:103
333. Hardy WB (1980) Dev Polym Photochem 3:287
334. Harper DJ, McKellar JF (1972) Chem Ind 4:848
335. Harper DJ, McKellar JF (1973) J Appl Polym Sci 17:3503
336. Henman TJ (1985) Dev Polym Degrad 6:107
337. Hiatt R, Clipsholm J, Vissar T (1964) Can J Chem 42:2754
338. Hidaka H, Zhao J, Kitamura K, Nohara K, Serpone N, Pelizzetti E (1992) J Photochem Photobiol A Chem 64:103
339. Hiraoka H, Chunang TJ, Masuhara M (1988) J Vac Sci Technol B6:463
340. Hiraoka H, Lazare S (1990) Appl Surf Sci 46:264
341. Hiraoka H, Lazare S, Cros A (1992) J Photochem Photobiol A Chem 65:293
342. Hirayama S, Foster RJ, Mellor JM, Whitling PH, Grant KR, Philips D (1978) Europ Polym J 14:679
343. Holden DA (1987) Encycl Polym Sci Eng 11:154
344. Holländer A, Klemberg-Sapiecha JE, Wertheimer MR (1994) Macromolecules 27:2893
345. Hong SI (1990) Makromol Chem Makromol Symp 33:213
346. Hoyle CE, Kim KJ (1986) J Appl Polym Sci 24:1879
347. Hoyle CE, Kim KJ (1987) J Polym Sci Chem Ed 25:2631
348. Hrdlovič P, Taimr L, Pospišil J (1989) Polym Degrad Stabil 25:73
349. Humphrey JS Jr, Roller RS (1971) Mol Photochem 3:35
350. Humphrey JS Jr, Shultz AR, Jacquiss DBG (1973) Macromolcules, 6:305
351. Huvet A, Phillipe J, Verdu J (1978) Europ Polym J 14:709
352. Ikeda T, Kawaguchi K, Yamaoka H, Okamura S (1978) Macromolecules 11:735
353. Inokuti M, Hirayama F (1965) J Chem Phys 43:1978
354. Irving M, Foldes E, Barabsa K, Kelen K, Tüdös F (1986) Polym Degrad Stabil 14:319
355. Ito M, Porter RS (1982) J Appl Polym Sci 27:4471
356. Iu KK, Ogilby PR (1987) J Phys Chem 91:1611, erratum (1988) J Phys Chem 92:5854
357. Iu KK, Ogilby PR (1988) J Phys Chem 92:4662
358. Ivanov VB, Zhuravlev MA (1986) Polym Photochem 7:55
359. Jacques PPL, Poller RC (1993) Europ Polym J 29:75
360. Jacques PPL, Poller RC (1993) Europ Polym J 29:83
361. Jellinek HHG (1967) J Appl Polym Sci Appl Polym Symp 4:41
362. Jellinek HHG (1973) Text Res J 43:557
363. Jellinek HHG (ed) (1979) Aspects of Degradation and Stabilization of Polymers, Elsevier, Amsterdam
364. Jellinek HHG (ed) Degradation and Stabilization of Polymers, Elsevier, Amsterdam, vol 1
365. Jellinek HHG, Chaudhuri AK (1972) J Polym Sci Chem Ed 10:1773
366. Jellinek HHG, Flajsman F (1969) J Polym Sci Chem Ed 7:1153
367. Jellinek HHG, Flejsman F, Kryman FJ (1969) J Appl Polym Sci 13:107
368. Jellinek HHG, Hrdlovič P (1971) J Polym Sci Chem Ed 9:1219
369. Jellinek HHG, Kachi H (eds) (1989) Degradation and Stabilization of Polymers, Elsevier, Amsterdam, vol 2
370. Jellinek HHG, Kryman FJ (1969) J Appl Polym Sci 13:2504
371. Jellinek HHG, Lipovač SN (1970) Macromolecules 3:237
372. Jellinek HHG, Pavlineč J (1970) In: Photochemistry of Macromolecules, Plenum Press, New York, p 91
373. Jellinek HHG, Schlueter WA (1960) J Appl Polym Sci 3:206
374. Jellinek HHG, Srinivasan R (1984) J Phys Chem 88:3048
375. Jellinek HHG, Toyoshima Y (1967) J Polym Sci Chem Ed 5:3214
376. Jellinek HHG, Wang LC (1965) Kolloid Z Z Polym 202:1
377. Johns HE (1971) In: Creation and Detection of the Excited States (Lamola AA, ed) Dekker, New York, vol 1A, p 123
378. Johnson R (1988) J Plast Film Sheeting 4:155
379. Jörg F, Schmitt D, Ziegahn KF (1984) Staub Reinhalt Luft 44:276
380. Jouan KT, Gardette JL (1987) Polym Commun 28:239
381. Jouan X, Adam C, Fromaget D, Gardette JL, Lemaire J (1989) Polym Degrad Stabil 25:247
382. Kaczmarek H, Lindén LÅ, Rabek JF (1995) J Polym Sci Chem Ed 33:879

383. Kaczmarek H, Lindén LÅ, Rabek JF (1995) Polym Degrad Stabil 37:33
384. Kagyia T, Nishimoto S, Watanabe Y, Kato M (1985) Polym Degrad Stabil 12:261
385. Kalmus CE, Hercules DM (1974) J Amer Chem Soc 96:449
386. Kaplan ML, Kelleher PG (1970) J Polym Sci Chem Ed 8:3163
387. Kaplan ML, Trozzolo AM (1979) In: Singlet Oxygen (Wasserman HH, Murray RW, eds) Academic Press, p 575
388. Kato K, Okamura S, Yamaoka H (1976) J Polym Sci Polym Lett 14:211
389. Kato K, Sasaki H, Okamura S (1980) J Polym Sci Polym Lett 18:197
390. Kawaoka K, Khan AU, Kearne DR (1967) J Chem Phys 46:1842
391. Kawaoka K, Khan AU, Kearns DR (1967) J Chem Phys 47:1883
392. Kearns DR (1971) Prepr Div Petrol Chem Amer Chem Soc 16:A9
393. Kearns DR (1971) Chem Rev 71:395
394. Kearns DR, Hollins RA, Khan AU, Chambers RW, Radlick P (1971) J Amer Chem Soc 89:5455
395. Kearns DR, Khan AU (1969) Photochem Photobiol 10:193
396. Kelen T (1983) Polymer Degradation,Van Nostrand Reinhold, New York
397. Khan AU, Kasha M (1968) Adv Chem Ser 77:143
398. Khan AU, Kearms DR (1968) J Chem Phys 48:3272
399. Kiefer RL, Orwoll RA (1989) Sci Techol Aerosp Res 27:1
400. Kilp T, Guillet JE, Merle-Aubry L, Merle Y (1982) Macromolecules 15:60
401. Kinard WH, Martin GD (1991) Proceedings of NASA LDEF 69 Months in Space Symposium, NASA, Part 1, p 49
402. Kiryushkin SG,Yakimchenko OE, Shlyapintokh YuA, Pariskii GB, Toptygin GYa, Lebedev YaS (1975) Vysokomol Soedin A 17:738
403. Klemchuk PP (1990) Polym Degrad Stabil 27:183
404. Klevens HB (1952) J Polym Sci 10:97
405. Klöpffer W (1991) EPA Newslett 41:24
406. Kockott D (1989) Polym Degrad Stabil 25:181
407. Koenig JL (1983) Adv Polym Sci 54:87
408. Koenig T (1969) J Amer Chem Soc 91:255
409. Koizumi M, Kato S, Mataga N, Matsuura T, Usui Y (1978) Photosensitized Reactions, Kagakudojin Publ, Kyoto
410. Koren G (1984) Appl Phys Lett 45:10
411. Koren G, Yeh JTC (1984) Appl Phys Lett 44:1112
412. Kramer EJ (1983) Adv Polym Sci 52/53:1
413. Kramer EJ, Berger LL (1990) Adv Polym Sci 91/92:1
414. Kristiansen M, Scurlock RD, Iu KK, Ogilby PR (1991) J Phys Chem 95:5190
415. Kryszewski M, Galeski A, Milczarek P (1976) J Polym Sci Polym Lett 14:365
416. Kubota H, Ogiwara Y (1989) Polym Degrad Stabil 29:207
417. Kubota H, Takahashi K, Ogiwara Y (1989) Polym Degrad Stabil 23:201
418. Kubota H, Takahashi K, Ogiwara Y (1990) Polym Degrad Stabil 29:207
419. Kubota H, Takahashi K, Ogiwara Y (1991) Polym Degrad Stabil 33:115
420. Kumakura A, Kobayashi K, Uchida Y, Osawa Z (1992) Polym Degrad Stabil 35:283
421. Kurata M, Tsunashima Y (1989) In: Polymer Handbook (Brandrup J Immergut EH, eds) Wiley, New York, p VII–1
422. Kuroda S, Nagura A, Horie K, Mita I (1989) Europ Polym J 25:621
423. Kurusu Y, Nishiyama H, Okawara M (1967) J Chem Soc Japan Ind Chem Sect 70:593
424. Kuzina SI, Mikhailov AI (1993) Europ Polym J 29:1589
425. Kysel O, Chmela S, Hlouskova Z (1993) Makromol Chem Rapid Commun 14:45
426. Lacoste J, Carlsson DJ (1992) J Polym Sci Chem Ed 30:493
427. Laist D (1987) Marine Pollut Bull 18:319
428. Lala D, Rabek JF (1981) Europ Polym J 12:7
429. La Mantia FP, Spadaro G, Acierno D (1985) Polym Photochem 6:425
430. Lawrence JB, Weir NA (1973) J Polym Sci Chem Ed 11:105
431. Lawton EL (1972) J Polym Sci Chem Ed 10:1857
432. Lazare S, Granier V (1988) Makromol Chem Makromol Symp 18:193
433. Lazare S, Granier V (1989) Laser Chem, 10:25
434. Lazare S, Srinivasan R (1986) J Phys Chem 90: 2124
435. Leaver IH (1980) In: Photochemistry of Dyed and Pigmented Polymers (Allen NS, McKellar JF, eds) Applied Science, London, p 161
436. Leaversuch R (1987) Mod Plast Intern 17:94
437. Leger LJ (1982) NASA Technical Memorandum TM-58246

438. Leger LJ, Visentine JT (1986) J Spacecr Rockets, 23:505
439. Leger L, Visentine JT, Santos-Mason B (1986) Int SAMPE Techn Conf 18:1015
440. Leidner J (1981) Plaste Waste, Dekker, New York
441. Leighton PA (1961) Photochemistry of Air Pollution, Academic Press, New York
442. Lemaire J, Arnaud R (1984) Polym Photochem 5:243
443. Lemaire J, Arnaud R, Lacoste J (1988) Acta Polym 39:27
444. Lemaire J, Gardette JL, Rivaton A, Roger A (1986) Polym Degrad Stabil 15:1
445. Lemoine P, Blau W, Dury A, Keely C (1993) Polymer, 34:5020
446. Letton A, Rock NI (1991) Proceedings of NASA LDEF 69 Months in Space Symposium, NASA, Part 2, p 705
447. Li SKL, Guillet JE (1984) Macromolecules 17:41
448. Li Bassi G (1986) Pitture e Vernici 62:30
449. Li Bassi G, Cadona L, Broggi F (1987) Proceedings of Radcure Europe'87 SME Edn, Sect 3–15
450. Lin CS, Liu WL, Chiu YS, Ho SY (1992) Polym Degrad Stabil 38:125
451. Lin SH (1980) Radiationless Transitions, Academic Press, New York
452. Lindén LÅ, Rabek JF, Kaczmarek H, Kaminska A, Scoponi M (1993) Coord Chem Rev 125:195
453. Lindenmeyer PH (?) Principles of Nonlinear Irreversible Thermodynamics Applied to the Testing of Materials, D-180-25583-1, Boening Co, Seattle
454. Linsker R, Srinivasan R, Wynne JJ, Alonso DR (1984) Lasers Surg Med 4:201
455. Lissi EA, Encinas MV (1991) In: Photochemistry and Photophysics (Rabek JF, ed) CRC Press, Boca Raton, vol IV, p 221
456. Lo J, Lee SN, Pearce EM (1984) J Appl Polym Sci 29:35
457. Lorand JP (1972) Progr Inorg Chem 17:207
458. Lucas PC, Porter RS (1988) Polym Degrad Stabil 22:175
459. Lucki J, Rånby B (1979) Polym Degrad Stabil 1:1
460. Lucki J, Rånby B (1979) Polym Degrad Stabil 1:251
461. Lucki J, Rånby B, Rabek JF (1979) Europ Polym J 15:1101
462. Lukač I, Hrdlovič P, Manaček Z, Belluš D (1971) J Polym Sci Chem Ed 9:69
463. MacCallum JR (1966) Europ Polym J 2:413
464. MacCallum JR (1977) Dev Polym Degrad 1:237
465. MacCallum JR (1984) Macromol Chem (London) 3:331
466. MacCallum JR (1985) Dev Polym Degrad 6:191
467. MacCallum JR (1989) In: Comprehensive Polymer Science (Allen G, Bevington JC, eds) Pergamon Press, New York, vol 6, p 529
468. MacCallum JR, Ramsay DA (1977) Europ Polym J 13:945
469. MacCallum JR, Rankin CT (1971) J Polym Sci Polym Lett 9:751
470. MacCallum JR, Rankin CT (1974) Makromol Chem 175:2477
471. Macedo PB, Litovsky TA (1965) J Chem Phys 42:245
472. Maerov SB (1965) J Polym Sci Chem Ed 3:487
473. Mahan GD, Cole H, Liu YS, Philips HR (1988) Appl Phys Lett 53:2377
474. Mandelsohn MA, Navish FW Jr, Luck RM, Yeoman FA (1983) ACS Symp Ser 220:39
475. Marcotte FB, Campbell D, Cleaveland JA, Turner DT (1967) J Polym Sci Chem Ed 5:481
476. Marcus RA (1956) J Chem Phys 24:966
477. Marcus RA (1960) Faraday Discuss Chem Soc 29:21
478. Marcus RA (1964) Ann Rev Phys Chem 15:155
479. Marcus RA (1965) J Chem Phys 43:679
480. Marcus RA (1982) Faraday Discuss Chem Soc 74:7
481. Marcus RA, Sutin N (1985) Biochem Biophys Acta 811:265
482. Marcus RA, Sutin N (1986) Comm Inorg Chem 5:119
483. Masuhara H, Hiroka H, Domen K (1987) Macromolecules, 20:450
484. Mateo JL, Catalina F, Sastre R (1989) Rev Plast Mod 57:73
485. Mathur A, Malhur GN (1977) Popular Plastics (Nov) 17
486. Matsuura T (1977) Oxygen Oxidation-Chemistry and Active Oxygen, Maruzen, Japan
487. May SA, Fuentes EC, Sato N (1991) Polym Degrad Stabil 32:357
488. Mayo RR, Niki E, Decker C, Richardson H (1973) Polym Preprints, 14:1186
489. Maxim LD, Kuist CH, Meyer ME (1968) Macromolecules 1:86
490. McCargo M, Dammann RA, Cummings T, Carpenter C (1985) Eur Space Agency Spec Publ ESA SP-232, 91
491. McGlynn SP, Azumi T, Kinoshita M (1969) Molecular Spectroscopy of Triplet State, Prentice-Hall, Englewood Cliffs, New Jersey

492. McKellar JF, Allen NS (1979) Photochemistry of Man-Made Polymers, Applied Science, London
493. McNeil IC, Zulfiqar S (1985) Polym Degrad Stabil 10:189
494. Meaker T (1991) In: Spacecraft Systems Engineering (Fortescue PW, Stark JPW, eds) Wiley, Chichester p 393
495. Meier IK, Langsam M (1993) J Polym Sci Chem Ed 31:83
496. Melchore JA (1962) Ind Eng Chem Prod Res Develop 1: 232
497. Mellor DC, Moir AB, Scott G (1973) Europ Polym J 9:219
498. Mendelson MA, Navish FW Jr, Luck RM, Yeoman FA (1983) ACS Symp Ser 220: 39
499. Merlin A, Fouassier JP (1980) Angew Makromol Chem 86:109
500. Merlin A, Fouassier JP (1982) Makromol Chem 108:185
501. Michaels AS, Bixer HJ (1961) J Polym Sci 50:413
502. Mimura Y, Ohkubo T, Takeuchi T, Sekihava K (1978) Jpn J Appl Phys 17:541
503. Mita I, Takagi T, Horie K, Shindo Y (1984) Macromolecules 17:2256
504. Mopper K, Zhou X (1990) Science 250:661
505. Murthy AS, Rao CNR (1968) Appl Spectr Rev 2:69
506. Naito I, Tashiro K, Kinoshita A, Schnabel W (1983) J Photochem 23:73
507. Narisawa I, Ishikawa M (1990) Adv Polym Sci 91/92:353
508. Neiman MB (1965) Ageing and Stabilization of Polymers, Consultants Bureau, New York
509. Nemzek TL, Guillet JE (1977) Macromolecules 10:94
510. Newton MD, Sutin N (1984) Ann Rev Phys Chem 35:437
511. Ng HC, Guillet JE (1978) Macromolecules 11:929
512. Ng HC, Guillet JE (1978) Macromolecules 11:937
513. Ng HC, Guillet JE (1978) Photochem Photobiol 28:571
514. Ng HC, Guillet JE (1978) In: Singlet Oxygen. Reactions with Organic Compounds and Polymers (Rånby B, Rabek JF, eds) Wiley, Chichester p 278
515. Ng HC, Guillet JE (1985) Macromolecules 18:2294 and 2299
516. Nguen TL, Rogers CE (1985) Polym Mater Sci Eng 52:192
517. Nguen TQ (1994) Polym Degrad Stabil 46:99
518. Nielsen SE (1983) Polym Testing 3:303
519. Nielson RDM, Soutar I, Steedman N (1977) J Polym Sci Phys Ed 15:617
520. Niino H, Kawabata Y, Yabe A (1989) Japn J Appl Phys 28:L2225
521. Niino H, Nakano M, Nagano S, Nitta H, Yano K, Yabe A (1990) J Photopolym Sci Technol 3:53
522. Niino H, Nakano M, Nagano S, Yabe A, Miki T (1989) Appl Phys Lett 55:510
523. Niino H, Nakano M, Nagano S, Yabe A, Moriya H, Miki T (1989) J Photopolym Sci Technol 2:133
524. Niino H, Ohana T, Ouchi A, Yabe A (1992) J Photopolym Sci Technol 5:301
525. Niino H, Shimoyama M, Yabe A (1990) Appl Phys Lett 57:2368
526. Niino H, Yabe A (1992) J Photochem Photobiol A Chem 65:303
527. Niino K, Yabe A (1992) Appl Phys Lett 60:2697
528. Niino H, Yabe A, Nagano S, Miki T (1989) Appl Phys Lett 54:2159
529. Novis Y, De Meulemeester R, Chtaib M, Pireaux JJ, Caudano R (1989) Brit Polym J 21:147
530. Novis Y, Pireaux JJ, Brezini A, Petit E, Caudano R, Lutgen P, Feyder G, Lazare S (1988) J Appl Phys 64:365
531. Nowakowska M, Kowal J, Waligura B (1978) Polymer 19:1317
532. Nowakowska M, Najbar J, Waligura B (1975) Europ Polym J 12:387
533. Noyes RM (1956) J Amer Chem Soc 78:5486
534. Noyes RM (1959) J Amer Chem Soc 81:566
535. Noyes RM (1961) In: Progress in Reaction Kinetics (Porter G, ed) Pergamon Press, New York, p 129
536. Nuzzo RG, Smolinksy G (1984) Macromolecules 17:1013
537. O'Brien RJ, Hard TM (1993) ACS Symp Ser 232:323
538. Occhiello E, Garbassi F, Malatesta V (1989) Angew Makromol Chem 169:143
539. O'Donnell B, White JR (1994) Polym Degrad Stabil 44:211
540. O'Donnell B, White JR (1995) J Mater Sci (in press)
541. O'Donnell B, White JR, Holding SR (1994) J Appl Polym Sci 52:1607
542. Ogilby PR, Dillon MP, Gao Y, Iu KK, Kristiansen M, Taylor VL, Clough RL (1993) ACS Symp Ser 236:573
543. Ogilby PR, Kristiansen M, Clough RL (1990) Macromolecules, 23:2698
544. Ogilby PR, Sanetra J (1993) J Phys Chem 97:4689

545. Oldershaw GA (1993) In: Photochemistry and Polymeric Systems (Kelly JM, McArdle CB, de F Maunder MJ eds) Royal Chem Soc Ed p 54
546. Olea AF, Encinas MV, Lissi EA (1982) Macromolecules 15:1111
547. Omichi H (1983) In: Degradation and Stabilization of Polyolefins (Allen NS, ed) Elsevier, London, p 187
548. O'Neal RL, Lightner EB (1991) Proceedings of NASA LDEF 69 Months in Space Symposium, NASA, Part 1, p 3
549. Osawa Z (1986) Photodegradation and Stabilization of Polymers: Funda-mentals and Practice of Photostabilization Techniques, CMC Co, Japan
550. Osawa Z, Kobayashi K (1985) J Polym Sci Polym Lett 23:41
551. Osawa Z, Kuroda H (1982) J Polym Sci Polym Lett 20:577
552. Osawa Z, Moriyama C, Nakano H (1981) J Polym Sci Chem Ed 19:1877
553. Osawa Z, Nakano H, Mitsui E, Nakano M (1979) J Polym Sci Chem Ed 17:139
554. Osawa Z, Uchida Y (1982) J Polym Sci Chem Ed 20:2259
555. Osborne KR (1959) J Polym Sci 38:357
556. O'Shaugnessy B (1991) J Chem Phys 94:4042
557. Oster G, Oster GA, Moroson H (1959) J Polym Sci 34:671
558. Owen ED, Bailey RJ (1972) J Polym Sci Chem Ed 10:113
559. Owen ED, Pasha I (1980) J Appl Polym Sci 25:2417
560. Owen ED, Williams JI (1973) J Polym Sci Chem Ed 11:905
561. Pabiot J, Verdu J (1981) Polym Eng Sci 21:32
562. Panaksem S, Kuczynski J, Thomas JK (1994) Macromolecules 27:3773
563. Papet G (1987) Radiat Phys Chem 29:65
564. Partridge RH (1967) J Chem Phys 47:4223
565. Patsis AD (1989) Advances in Stabilization and Controlled Degradation of Polymers, Technomic, Lancaster
566. Pavia FL, Lapman GM, Kriz GS (1982) In: Introduction to Organic Techniques: A Contemporary Approach, 2nd edn. Saunder Publ p 410
567. Peeling J (1983) J Polym Sci Chem Ed 21:2047
568. Peeling J, Clark DT (1981) Polym Degrad Stabil 3:97
569. Pegram JE, Andrady AL (1989) Polym Degrad Stabil 26:333
570. Perrin F (1924) C R Acad Sci 178:1978
571. Philips D, Roberts AJ, Soutar I (1983) Macromolecules 16:1593
572. Phillips GO, Worthington NW, McKellar JF, Sharpe RR (1967) Chem Commun 835
573. Phillips GO, Worthington NW, McKellar JF, Sharpe RR (1969) J Chem Soc A 767
574. Pippin HG (1989) Surface Coat Technol 39/40:595
575. Plummer CJ, Donald AM (1989) J Polym Sci Phys Ed 27:325
576. Popov A, Rapoport N, Zaikov G (1991) Oxidation of Stressed Polymers, Gordon & Breach Science Publ, New York
577. Porter CFC, Philips D (1994) Europ Polym J 30:189
578. Potter T, Schmelzer H, Baker R (1984) Progr Org Coat 12:321
579. Powell JG, Ellis IA (1987) Proceedings of IV International Conference on Lasers in Manufacturing, p 69
580. Pratte JF, Webber SE (1982) Macromolecules 15:417
581. Price TR, Fox RB (1966) J Polym Sci Polym Lett 4:771
582. Pringsheim J, Vogel M (1946) Luminescence of Fluids and Solids and its Practical Applications, Interscience, New York
583. Pruter A (1987) Marine Poll Bull 18:305
584. Pryde CA (1985) ACS Symp Ser 280:329
585. Quayyum MM, White JR (1985) J Mater Sci 20:2558
586. Quayyum MM, White JR (1993) Polym Degrad Stabil 41:163
587. Qu BJ, Xu YA, Shi WF, Rånby B (1992) Macromolecules 25:5215
588. Qu BJ, Xu YH, Shi WF, Rånby B (1992) Macromolecules 25:5220
589. Raab M, Hnát V, Schmidt P, Kotulák L, Tamir L, Pospisil J (1987) Polym Degrad Stabil 18:123
590. Raab M, Kotulák L, Kolařik J, Pospišil J (1982) J Appl Polym Sci 27:2457
591. Raab M, La Mantia F, Pospišil J (1990) Angew Makromol Chem 176/177:93
592. Rabek JF (1964) J Polym Sci Polym Lett 2:577
593. Rabek JF (1965) J Appl Polym Sci 9:2121
594. Rabek JF (1966) Wiad Chem 20:291, 355,435
595. Rabek JF (1967) J Polym Sci Polym Symp 16:949
596. Rabek JF (1968) Photochem Photobiol 7:5

597. Rabek JF (1971) Proceedings 3rd International Congress IUPAC, Butterworths, London, vol 8, p A90
598. Rabek JF (1973) Proceedings Conference on Photopolymers: Principles, Processes and Materials, Soc Plast Eng Mid-Hudson, p 27
599. Rabek JF (1975) In: Degradation of Polymers (Bamford CH, Tipper CFH, eds) Comprehensive Chemical Kinetics, Elsevier, Amsterdam, vol 14, p 425
600. Rabek JF (1976) ACS Symp Ser 25:255
601. Rabek JF (1980) Experimental Methods in Polymer Chemistry, Wiley, Chichester
602. Rabek JF (1982) Experimental Methods in Photochemistry and Photophysics, Wiley, Chichester
603. Rabek JF (1984) In: Polymer Additives (Kresta JE, ed) Plenum, New York, p 1
604. Rabek JF (1985) In: Singlet Oxygen (Frimer AA, ed) CRC Press, Boca Raton, vol 4, p 1
605. Rabek JF (1985) In: New Trends in Photochemistry of Polymers (Allen NS, Rabek JF, eds) Elsevier Applied Science, London, p 265
606. Rabek JF (1987) Mechanisms of Photophysical Processes and Photochemical Reactions in Polymers: Theory and Applications, Wiley, Chichester
607. Rabek JF (1988) Prog Polym Sci 13:83
608. Rabek JF (1990) Photostabilization of Polymers: Principles and Applications, Elsevier Applied Science, London
609. Rabek JF (1995) Photodegradation of Polymers: Mechanisms and Experimental Methods, Chapman & Hall, London
610. Rabek JF, Canbäck G, Lucki J, Rånby B (1976) J Polym Sci Chem Ed 14:1447
611. Rabek JF, Canbäck G, Rånby B (1979) J Appl Polym Sci Appl Polym Symp 35:299
612. Rabek JF, Lindén LA, Kaczmarek H, Qu BJ, Shi WF (1992) Polym Degrad Stabil 37:33
613. Rabek JF, Lucki J, Rånby B (1979) Europ Polym J 15:1089
614. Rabek JF, Rabek T (1963) J Appl Polym Sci 7:S38
615. Rabek JF, Rånby B (1973) In: ESR Applications to Polymer Research, Nobel Symposium 22 (Kinell PO, Rånby B, eds) Almqvist & Wiksel, Stockholm, p 201
616. Rabek JF, Rånby B (1974) J Polym Sci Chem Ed 12:273
617. Rabek JF, Rånby B (1974) J Polym Sci Chem Ed 12:295
618. Rabek JF, Rånby B (1974) Proceedings of International Symposium Degradation and Stability of Polymers, Brussels, p 257
619. Rabek JF, Rånby B (1975) Polym Eng Sci 15:40
620. Rabek JF, Rånby B (1976) J Polym Sci Chem Ed 14:1463
621. Rabek JF, Rånby B (1978) Photochem Photobiol 28:557
622. Rabek JF, Rånby B (1979) Photochem Photobiol 30:133
623. Rabek JF, Rånby B (1979) J Appl Polym Sci 23:2481
624. Rabek JF, Rånby B (1980) Rev Roum Chim 25:1045
625. Rabek JF, Rånby B, Arct J, Liu R (1984) J Photochem 25:519
626. Rabek JF, Rånby B, Östensson B, Flodin P (1979) J Appl Polym Sci 24:2407
627. Rabek JF, Rånby B, Skowronski TA (1985) Macromolecules, 18:1810
628. Rabek JF, Sanetra J (1986) Macromolecules 19:1679
629. Rabek JF, Sanetra B, Rånby B (1986) Macromolecules 19:1674
630. Rabek JF, Shur YJ, Rånby B (1978) In: Singlet Oxygen: Reactions with Organic Compounds and Polymers (Rånby B, Rabek JF, eds) Wiley, Chichester p.264
631. Ram A, Zilber O, Kening S (1985) Polym Eng Sci 25:535
632. Rånby B, Rabek JF (1975) Photodegradation, Photooxidation and Photostabilization of Polymers: Principles and Applications, Wiley, London
633. Rånby B, Rabek JF (1976) ACS Symp Ser 25:391
634. Rånby B, Rabek JF (1977) ESR Spectroscopy in Polymer Research, Springer, New York
635. Rånby B, Rabek JF (eds) (1978) Singlet Oxygen: Reactions with Organic Compounds and Polymers, Wiley, Chichester
636. Rånby B, Rabek JF (1979) J Appl Polym Sci 35:243
637. Rånby B, Rabek JF (1983) ACS Symp Ser 229:291
638. Rånby B, Rabek JF (1992) In: Comprehensive Polymer Science (Allen G, ed) Pergamon Press, Oxford, p 253
639. Rånby B, Rabek JF, Joffe Z (1973) Proceedings of Conference Degradability of Polymers and Plastics, Institution of Electrical Engineers, London, p 3/1
640. Rasool SI (1973) Chemistry of Lower Atmosphere, Plenum Press, New York
641. Redpath AE (1987) Symposium on Degradable Plastics, SPI Washington
642. Rehorek D, DiMartino S, Kempt TJ, Hennig H (1989) J Inf Rec Mat 17:469
643. Rehorek D, Henning H (1982) Can J Chem 60:1565

644. Rehorek D, Janzen EG (1986) J Photochem 35:251
645. Reichmanis E, Wilkins CW Jr (1980) Org Coat Plast Chem 43:243
646. Reichmanis E, Wilkins CW Jr, Chandros EA (1980) J Electrochem Soc 127:2514
647. Reiser A (1989) Photoreactive Polymers: The Science and Technology, Wiley, New York
648. Reiser A, Edgerton PL (1979) Macromolecules 12:670
649. Reneker DH, Bolz LH (1976) J Makromol Chem Sci Chem A10:599
650. Rice S, Jain K (1984) Appl Phys A33:195
651. Rigaudy J, Klesney SP (1979) Nomenclature of Organic Chemistry, Pergamon Press, Oxford
652. Rivaton A, Morel P (1992) Polym Degrad Stabil, 35:3
653. Rivaton A, Sallet S, Lemaire J (1983) Polym Photochem 3:463
654. Rivaton A, Sallet D, Lemaire J (1986) Polym Degrad Stabil 14:1
655. Rivaton A, Sallet D, Lemaire J (1986) Polym Degrad Stabil 14:23
656. Robey MJ, Field G, Styzinski M (1989) Mater Forum 13:1
657. Roginsky VA (1984) Develop Polym Degrad 5:193
658. Rosato DV (1968) In: Environmental Effects on Polymeric Materials (Rosato DV, Schwartz RT, eds) Wiley, New York, p 733
659. Rosato DV, Schwartz RT (eds) Environmental Effects on Polymeric Materials, Wiley, New York
660. Rozantsev EG (1970) Free Nitroxyl Radicals, Plenum Press, New York
661. Rychly J, Rychla L, Klimova M (1993) Polymer 34:4961
662. Salomon GA, Noyes RM (1962) J Amer Chem Soc 84:672
663. Schoolenberg GE (1988) J Mater Sci 23:1580
664. Schoolenberg GE, Meijer HDF (1991) Polymer 32:438
665. Schoolenberg GE, Vink P (1991) Polymer 32:432
666. Scott G (1973) J Oil Colour Chem Assoc 56:521
667. Scott G (1973) In: Polymers and Ecological Problems (Guillet JE, ed) Plenum Press, New York
668. Scott G (1973) Macromol Chem 8:319
669. Scott G (1976) Resources, Recovery and Conserv 1:381
670. Scott G (1976) J Polym Sci Polym Symp 57:357
671. Scott G (1978) In: Singlet Oxygen: Reactions with Organic Compounds and Polymers (Rånby B, Rabek JF, eds) Wiley, Chichester, p 230
672. Scott G (1984) J Photochem 25:83
673. Scott G (1985) Polym Degrad Stabil 10:97
674. Scott G (1989) Polym News 14:169
675. Scott G (ed) (1990) Mechanisms of Polymer Degradation and Stabilization, Elsevier Applied Science, London
676. Scott G (1990) Polym Degrad Stabil 29:135
677. Scott G (1990) J Photochem Photobiol AChem 51:73
678. Scurlock RD, Martire DO, Ogilby PR, Taylor VL, Clough RL (1994) Macromolecules 27:4787
679. Scurlock RD, Ogilby PR (1993) J Photochem Photobiol A Chem 72:1
680. Seguchi T, Tamura N (1973) J Phys Chem 77:40
681. Selke SEM (1990) Packaging and the Environmental Alternatives, Trends and Solution, Technomic Publication, Lancaster
682. Shah H, Rufus IB, Hoyle CE (1994) Macromolecules 27:553
683. Shard AG, Badyal JPS (1991) Polym Commun 32:217
684. Sheldrick GE, Vogl O (1976) Polym Eng Sci 16:65
685. Shelton JR, Kopazewski RF (1967) J Org Chem 32:2908
686. Sherman ES, Ram A, Kenig S (1982) Polym Eng Sci 22:457
687. Shibryaeva LS, Kiryushkin SG, Zaikov GE (1992) Polym Degrad Stabil 36:17
688. Shimada S, Kashiwabara H, Sohma K (1969) Rep Progr Polym Phys Japan 12:465
689. Shimada S, Kashiwabara H, Sohma J (1970) J Polym Sci Phys Ed 8:1291
690. Shimada S, Kashiwabara H, Sohma J (1971) Rep Progr Polym Phys Japan 14:547
691. Shimoyama M, Niino H, Yabe A (1992) Makromol Chem 193:569
692. Shlyapintokh VYa (1982) Dev Polym Photochem 3:215
693. Shlyapintokh VYa (1985) Photochemical Conversions and Polymer Stabilization, Hanser, München
694. Shlyapintokh VYa, Kiryushkin SG, Marin AP (1986) Antioxidative Stabilization of Polymers, Khimia, Moscow, p.50
695. Shultz AR (1958) J Chem Phys 29:200
696. Shultz AR (1960) J Polym Sci 47:267

697. Siegman A, Fauchet P (1986) IEEE J Quantum Electron 22:1384
698. Siesler HW (1984) Adv Polym Sci 65:1
699. Siesler HW (1993) ACS Symp Ser 236:41
700. Siesler HW, Holland-Moritz, K (1980) Infrared and Raman Spectroscopy of Polymers, Dekker, New York
701. Sikkema K, Cross GS, Hanner MJ, Priddy DB (1992) Polym Degrad Stabil 38:113
702. Singh RP, Mani R, Sivaram S, Lacoste J, Vaillant D, (1993) J Appl Polym Sci 50:1871
703. Skowronski AT, Rabek JF, Rånby B (1984) Polym Photochem 5:77
704. Smets GJ (1985) Polym J 17:153
705. Smoluchowski MW (1917) Z Phys Chem 92:192
706. Somersall AC, Guillet JE (1975) J Macromol Sci Rev Macromol Chem C, 135
707. Soutar I, Philips D, Roberts AJ, Rumbles G (1982) J Polym Sci Phys Ed 20:1759
708. Sperling LH (1986) Introduction to Physical Polymer Science, Wiley, New York
709. Srinivasan R (1983) J Radiat Curing Oct. 12
710. Srinivasan R (1983) J Vac Sci Techn B1:923
711. Srinivasan R (1986) Science 234:559
712. Srinivasan R (1987) Polym Degrad Stabil 17:193
713. Srinivasan R (1991) Appl Phys Lett 58:2895
714. Srinivasan R (1991) J Appl Phys 70:7588
715. Srinivasan R (1993) Polym Degrad Stabil 43:101
716. Srinivasan R (1993) In: Photochemistry and Polymeric Systems (Kelly JM, McArdle CB, de Maunder F, eds) Royal Chemical society, Cambridge, p 47
717. Srinivasan R, Braren B (1984) J Polym Sci Chem Ed 22:2601
718. Srinivasan R, Braren B (1988) Appl Phys A45:289
719. Srinivasan R, Braren B (1989) Chem Rev 89:1303
720. Srinivasan R, Braren B (1990) In: Lasers in Polymer Science and Technology (Fouassier JP, Rabek JF, eds) CRC Press, Boca Raton, vol 3, p 133
721. Srinivasan R, Braren B, Casey KG (1990) J Appl Phys 68:1842
722. Srinivasan R, Braren B, Dreyfus RW (1986) Appl Phys Lett 48:445
723. Srinivasan R, Braren B, Dreyfus RW, Hadel L, Seeger DE (1986) J Opt Soc Amer 3:785
724. Srinivasan R, Braren B, Seeger DE, Dreyfus RW (1986) Macromolecules 19:916
725. Srinivasan R, Casey K, Braren B (1989) Chemtronics 4:153
726. Srinivasan R, Dreyfus RW (1985) In: Laser Spectroscopy (Hanch TW, Shen YR, eds) Springer, New York, vol 1, p 396
727. Srinivasan R, Leigh W (1982) J Amer Chem Soc 104:6784
728. Srinivasan R, Mayne-Banton V (1982) Appl Phys Lett 41:576
729. Srinivasan R, Sutcliffe E, Braren B (1988) Laser Chem 9:147
730. Stark JPW (1991) In: Spacecraft Systems Engineering (Fortescue PW, Stark JPW eds) Wiley, Chichester, p 9
731. Stein BA, Young PR (1990) LDEF Materials Data Analysis, Workshop NASA CP-10046
732. Štěpek J, Duchaček V, Čurda D, Horaček J, Špivek M, Panchartek J, Reynolds GEJ (1987) Polymers as Materials for Packaging, Ellis Harwood, Chichester
733. Stewart LC, Carlsson DJ, Wiles DM, Scaiano JC (1983) J Amer Chem Soc 105:3605
734. Stokes S, Fox RB (1962) J Polym Sci 56:507
735. Ström G, Fredriksson M, Klason T (1988) J Colloid Interface Sci 123:324
736. Stryer L, Haugland RP (1967) Proc Nat Acad Sci US 58:719
737. Sugita K, Ueno N, Konishi S, Suzuki Y (1983) Photog Sci Eng 27:146
738. Sutcliffe E, Srinivasan R (1986) J Appl Phys 60:3315
739. Swain H, Silbert L, Miller J (1964) J Amer Chem Soc 86:2562
740. Tabata M, Sohma J, Yamaoka H, Matsuyama T (1985) Chem Phys Lett 119:256
741. Tabata M, Sohma J, Yamaoka H, Matsuyama T (1986) Radiat Phys Chem 27:35
742. Tagliaferri V, Crivelli Visconti I, Di Ilio A (1987) Proceedings of 6th International Conference on Composite Materials, p 1-190
743. Tagliaferri V, Di Ilio A, Criveli Visconti I (1985) Composites 16:317
744. Tanielian C, Chaineaux J (1978) J Photochem 9:19
745. Tanielian C, Chaineaux J (1979) J Polym Sci Chem Ed 17:715
746. Tanielian C, Chaineaux J (1980) Europ Polym J 16:619
747. Tanielian C, Mechin R (1989) J Photochem Photobiol A Chem 48:43
748. Tarte P (1952) J Chem Phys 20:1560
749. Taylor LH, Tobias JW (1977) J Appl Polym Sci 21:1273
750. Taylor LJ, Tobias JW (1981) J Appl Polym Sci 26:2917
751. Tidjani A, Arnaud R, Dasilva A (1983) J Appl Polym Sci 47:211

752. Tiňo J, Urban J, Klimo V (1989) Polymer 30:2136
753. Tiňo J, Urban J, Klimo V (1990) Chem Papers 44:711
754. Tiňo J, Urban J, Klimo V (1992) Chem Papers 46:154
755. Tonyali K, Jensen LC, Dickinson JT (1988) J Vac Sci Technol A6, 941
756. Torikai A, Asada S, Fueki K (1986) Polym Photochem 7:1
757. Torikai A, Mitsuoka T, Fueki K (1993) J Polym Sci Chem Ed 31:2785
758. Torikai A, Murata HS (1984) Polym Degrad Stabil 4:225
759. Torikai A, Shirakawa H, Nagaya S, Fueki K (1990) J Appl Polym Sci 40:1637
760. Torikai A, Takeuchi T, Fueki K (1983) Polym Photochem 3:307
761. Torikai A, Takeuchi A, Fueki K (1986) Polym Degrad Stabil 14:367
762. Torikai A, Takeuchi A, Nagaya S, Fueki K (1986) Polym Photochem 7:199
763. Torre LP, Pippin HG (1986) Int SAMPE Tech Conf 18:1086
764. Trefonas IIIP, West R, Miller RD, Hofer D (1983) J Polym Sci Polym Lett 21:823
765. Trozzolo AM, Winslow FH (1968) Macromolecules 1:98
766. Tsubomura H, Mulliken RS (1960) J Amer Chem Soc 82:5966
767. Tsuji K (1977) Polym Plast Technol Eng 9:1
768. Tsuji K, Seiki T (1971) J Polym Sci Chem Ed 9:3063
769. Tüdös F, Iring M (1981) Plaste Kautsch 8:421
770. Turro NJ (1965) Molecular Spectroscopy, Benjamin, New York
771. Turro NJ (1977) Pure Appl Chem 49:405
772. Turro NJ (1978) Modern Molecular Photochemistry, Benjamin Cummings, Menlo Park
773. Ueno N, Konishi S, Tanimoto K, Sugita K (1981) Japan J Appl Phys 20:709
774. Vaidergorin EYL, Marcondes MER, Toscano VG (1987) Polym Degrad Stabil 19:329
775. Valenza A, La Mantia FP (1987) Polym Degrad Stabil 19:135
776. Valenza A, La Mantia FP (1988) Polym Degrad Stabil 20:63
777. Van de Wel H, Vroonhoven FCBM, Lub J (1993) Polymer 34:2065
778. Van Cleave RA (1979) Characteristic of Laser Cutting Kevlar Laminates, DOE Rept. No BDX-613-2075
779. Van Cleave RA (1980) Laser Cutting Plastic Materials, DOE Rept. No BDX-613-2476
780. Van Cleave RA (1981) Laser Cutting Shapes in Plastics, DOE Rept. BDX-613-2727
781. Van Cleave RA (1986) Laser Cutting Plastics, DOE Rept. No BDX-613-2906
782. Victor JG, Torkelson JM (1987) Macromolecules 20:2241
783. Vinogradov SN, Linnell LH (1972) Hydrogen Bonding, Van Nostrand, Reinhold, New York
784. Voigt J (1966) Die Stabilisierung der Kunststoffe gegen Licht und Wärme, Springer, Berlin
785. Wagner PJ (1976) In: Triplet States III, Topics in Current Chemistry, Springer, Berlin, vol 66, p 1
786. Walte TR (1957) Phys, Rev 107:463
787. Webber SE, Swenberg CE (1980) Chem Phys 49:231
788. Wei K, Mani JC, Pitts JNRr (1967) J Amer Chem Soc 89:4225
789. Weir NA (1978) Europ Polym J 14:9
790. Weir NA (1982) Dev Polym Degrad 4:143
791. Weir NA, Ceccarelli A (1993) Polym Degrad Stabil 41:37
792. Weir NA, Ceccarelli A (1993) Polym Degrad Stabil 41:93
793. Weir NA, Ceccarelli A, Arct J (1993) Europ Polym J 29:737
794. Weir NA, Jones DA, Kutok P (1989) Polym Commun 30:201
795. Weir NA, Milkie TH (1978) Makromol Chem 179:1989
796. Welch AJ, Motamedi M, Restegar S, Le Carpentier GL, Jansen D (1991) Photochem Photobiol A Chem 53:815
797. Wells RK, Baydal JPS (1992) J Polym Sci Chem Ed 30:2677
798. Whitaker AF, Jang BZ (1993) J Appl Polym Sci 48:1341
799. Wilhelm C, Gardette JL (1994) J Appl Polym Sci 51:1411
800. Wilkins CW Jr, Reichmanis E, Chandros EA (1980) J Electrochem Soc 127:2510
801. Wilski H (1963) Kolloid Z Z Polym 188:4
802. Wilson JE (1954) J Chem Phys 22:334
803. Winnik MA, Li CK (1982) J Photochem 19:337
804. Winslow FH (1963) Chem Ind 533:1465
805. Winslow FH, Hawkins WL (1967) J Appl Polym Sci Appl Polym Symp 4:29
806. Witsmontski-Knittel T, Kilp T (1983) J Polym Sci Chem Ed 21:3209
807. Wood DGM, Kollman TM (1972) Chem Ind (London) 423
808. Wu SK, Rabek JF (1988) Polym Degrad Stabil 21:365
809. Wypych J (1990) Weathering Handbook, Chemtec, Toronto

810. Yakimtchenko OE, Gaponova IS, Goldberg VM, Parisky GB, Toptygin DYa, Lebedev YaS
      (1974) Izv Akad Nauk SSSR, Ser Khim 352
811. Yakimtchenko OE, Kirushkin SG, Pariskii GB, Toptygin Dya, Shlyapintokh DYa, Lebedev
      YaS (1975) Izv Akad Nauk SSSR, Ser Khim 2235
812. Yamaguchi K (1985) In: Singlet Oxygen (Frimer AA, ed) CRC Press, Boca Raton, vol 3, p
      119
813. Yamamoto N, Akaishi S, Tsubomura H (1972) Chem Phys Lett 15:459
814. Yamaoka H, Ikeda T, Okamura S (1977) Macromolecules 10:717
815. Yamauchi J, Ando H, Yamaoka A (1993) Makromol Chem Rapid Commun 14:13
816. Yeh JTC (1986) J Vac Sci Technol A4:653
817. Young PR, Slemp WS (1983) ACS Symp Ser 527:278
818. Young PR, Slemp WS (1991) Proceedings of NASA LDEF 69 Months in Space Symposium,
      NASA, Part 2, p 687
819. Young RJ (1981) Introduction to Polymers, Chapman & Hall, London, p 204
820. Xu FY, Chien CW (1993) Macromolecules 26:3485
821. Zahradnickova A, Sedlar J, Dastych D (1991) Polym Degrad Stabil 32:155
822. Zamfirova-Ivanova G, Raab M, Pelzbauer Z (1981) J Mater Sci 16:3381
823. Zamotaev PV (1989) Polym Degrad Stabil 26:65
824. Zamotaev PV, Litsov NJ, Kachan AA (1986) Polym Photochem 7:139
825. Zamotaev PV, Luzgarev SV (1989) Angew Makromol Chem 173:47
826. Zamotaev PV, Mityukhin O, Luzgarev SV (1992) Polym Degrad Stabil 35:195
827. Zepp RG, Baughman GL, Schlotzhauer PF (1981) Chemosphere 10:109
828. Zeep RG, Wolfe NL, Baughman GL, Hollis RC (1977) Nature 267:421
829. Ziegel KD, Eirich FR (1970) J Polym Sci Phys Ed 8:2015
830. Zimnick DG, Tennyson RC, Kok LJ, Maag CR (1985) Eur Space Agency, Spec Publ ESA
      SP-232, 81-9
831. Zollinger H (1991) Colour Chemistry, 2nd ed, VCH, Weinheim
832. Zweig A, Henderson WA (1975) J Polym Sci Chem Ed 13:717
833. US Standard Atmosphere (1976) NOAA, Washington.

Subject Index

Printing: COLOR-DRUCK DORFI GmbH, Berlin
Binding: Buchbinderei Lüderitz & Bauer, Berlin